Economic Growth and Development

Hasan Gürak

Economic Growth and Development

Theresa, Criticisms and an *Alternative* Growth Model

Bibliographic Information published by the Deutsche Nationalbibliothek
The Deutsche Nationalbibliothek lists this publication in
the Deutsche Nationalbibliografie; detailed bibliographic
data is available in the internet at http://dnb.d-nb.de.
Library of Congress Cataloging-in-Publication Data
Gürak, Hasan.
  Economic growth and development : theories, criticisms and an alternative growth model / Hasan Gürak.
     pages cm
  ISBN 978-3-631-66072-0
 1. Economic development. 2. Labor productivity. 3. Technological innovations. 4. Income distribution. I. Title.
  HD75.G87 2015
  338.9001--dc23
                          2015002904

Coverillustration:
© Olaf Gloeckler, Atelier Platen, Friedberg

ISBN 978-3-631-66072-0 (Print)
E-ISBN 978-3-653-05435-4 (E-Book)
DOI 10.3726/ 978-3-653-05435-4

© Peter Lang GmbH
Internationaler Verlag der Wissenschaften
Frankfurt am Main 2015
All rights reserved.
PL Academic Research is an Imprint of Peter Lang GmbH.

Peter Lang – Frankfurt am Main · Bern · Bruxelles · New York ·
Oxford · Warszawa · Wien

All parts of this publication are protected by copyright. Any
utilisation outside the strict limits of the copyright law, without
the permission of the publisher, is forbidden and liable to
prosecution. This applies in particular to reproductions,
translations, microfilming, and storage and processing in
electronic retrieval systems.

This publication has been peer reviewed.

www.peterlang.com

Foreword

Knowledge is increased through observation, research, experiment and mental effort. Science and technology can only make progress through "new ideas" e.g., "new knowledge" developed by the free mental powers of human beings.

A person (a researcher) seeking for new knowledge is not only responsible for human beings but also for all creatures and for nature itself. Her/his new ideas are expected to affect the course of society's present and future. Therefore, a researcher makes her/his best contribution by posing questions and by completing her/his work by providing answers that the researcher considers to be correct. The recipients of this new knowledge have to be open-minded, tolerant and unbiased.

New ideas in the social sciences, naturally including economics, have to fulfill four major conditions. They must be:

1- Logical.
2- Consistent.
3- Explanatory.
4- Testable.

There are many economic theories containing all of the above features; and most of them belong to the mainstream ideology, i.e. the neoclassical theories/models. As a matter of fact, at present, almost all economic theories/models are of Anglo-Saxon origin. Yet, these dominant economic theories/models are, to a large extent, a part of a "grand parable".

They are no doubt logical, consistent and even explanatory in many ways, but when tested for their relevance to real economic situations, they fail to correctly explain the normal economic transactions. In other words, those theories/models which are dominant today only are successful in explaining "a fictional world" and "fictional economic relationships" which are largely based upon unrealistic assumptions. The assumptions of "perfect competition" alone are sufficient to reject the neoclassical theories/models as highly fictitious and unrealistic. One does not have to be an expert in the field of economics to realize these shortcomings.

It is high time to produce "new and alternative" theories and models to replace the "parables" of these mainstream ideologies. I hope that this book will appeal to open minded economists the world over, as a constructive contribution for the further development of "new economic ideas".

I'm grateful to "Hakim John Lee" from Bodrum-Turgutreis for his contribution correcting my errors in English.

Hasan Gürak
2014

Contents

Foreword 5
Introduction 15
The Purpose & Scope of this Study 16
Which Criteria Can We Use as a Comparison? 18

Chapter 1: Growth Process Worldwide 21

All About Growth 25
"Savings = Investment" Paradox 27
On Added-value Criterion 30
The Relationship Between the Laborer – Technological Progress & Growth 31
What Role Does Investment Play in Regard
to Technology and Long-run Growth? 34
Access to and the Use of Technology 34
Laborer, Knowledge & Growth 36
Qualified Labor or "Human Capital" 37
Creative Labor and Growth 39
Unqualified Labor 39
Technological Progress & the Global Economy 39
Is it possible to Measure the "Qualification Level
of the Laborer" Accurately? 40
Other Factors Having an Impact on Growth 42
The Institutional and Cultural Infrastructure 42
Competitive Conditions 43
Economic and Political Stability 43
Financing the Investments 44
Natural Resources 44

Chapter 2: Basic Concepts Related to Growth 45

Productivity- (Productivity) Growth - Qualified
Laborer - Technological Progress 45
Productivity (V) 47
Measuring Productivity: a Static Analysis 48
The Quantitative Analysis 48

Measuring Qualitative Differences 49
Value Analysis of "Productivity" 49
An Alternative Value Measurement...50
Optimum Productivity 51
"(Productivity) Growth" 51
Quantitative Productivity Growth 52
Partial Factor Productivity Growth..53
Added-value Based Productivity Growth 54
Is a "Value" Criterion a Perfect Choice? 54
"New Products" and Growth 55
An Inter-Country Productivity Comparison 55
Concluding Remarks 56

Chapter 3: Growth Theories: Historical Perspective..............................57

A. Smith 60
Ricardo..63
Marx..65
Marshall..68
Keynes...70
Keynes's Model..71
The Characteristics of Keynes' "Static" Model 73
Harrod-Domar 74
A summary of the Harrod-Domar model..75
Schumpeter 75
Overview 76

Chapter 4: New Approaches to Growth Theory 79

Neoclassical Growth Theory-1: Pre-Solow..80
A Criticism of the Theory in Regard to Growth 85
Neoclassical Growth Theory-2: Solow & After 86
Estimation of TFP 88
Criticism of Solow's Model & the TFP Approach 89
TFP & Long-run Growth 90
Some TFP-related Data..94
TFP and Growth in Turkey 95
"Endogenous" Growth Models..96
Lucas: the Mechanical Model of Growth 97
The Institutional School 105

P. Romer: An Endogenous Growth Theory 109
R.J. Barro on Growth 120
Aghion-Howitt: Creative Destruction 123
Sectors and Production 126
Grossman-Helpman: Foreign Trade and Growth 128
G. Mankiw 135
Middle-Income-Trap: End of Growth? 139
In Conclusion 141
Concluding Remarks for the Chapter 144

Chapter 5: An Alternative Growth Model 145

Introduction 145
The Purpose 146
"Productive" Factors & "Production" Factors or Inputs 147
Productive Factors 148
Factors (Inputs) of Production 148
"Productive" Factors and Value-creation 148
The Genesis of Growth 149
Some Basic Assumptions 152
An Attempt at an Alternative Growth Model 153
Initial Case: A Simple "Stationary" Output,
Exchange & Distribution Model 153
A Simple "Short-run" Growth Model: 1 154
Assuming Access to a "New" Market 156
A Simple "Long-run" Growth Model: 2 157
"New Markets" for the "New Product" 158
Conlusions to be Drawn from the Simple Growth Models 159
Limit of Short-run Growth 159
Technological Innovations and Producers 160
The Importance of Innovations 161
Conclusions 163
A Note on Labor, Innovation & Value-Price Theory 163

Chapter 6: Short-run Growth in the Real Economy 165

1- Short-run Growth: A "Given" Technology 165
1-a) EE and/or TE growth 165
1-b) Extended Production for New Markets 167
Eight Methods to Increase Short-run Growth 168

Chapter 7: Long-run Groth in the Real Economy 177

Technological (Macro) Productivity Growth 177
a) A "Given" Product, but a "New" Production Method 177
b-) New Products 179
Growth: Reconsidered - Both in the Short- & Long-run 181
The Concrete Effects of Innovations 182
Given Product – New Method of Production 182
Technological Productivity Growth 184
"Given" Product – "New" Production Process 184
New Product 189
New Product and Functional Income Distribution 190
New Product, "Monopoly" and Profit Rate 191
Innovations, Growth and Price 192
A "Given Product" but a "New Production Process" 192
"New" Product and "New" Price 193
Other Factors Influencing Price Level 193
Technology Transfer & Long-run Growth 195
Measures to be Taken 196
Major Costs of Technology Transfer Through FDIs 198
Conclusion 199

Chapter 8: Growth & The Service Sector 203

"Productivity" & "Productivity Growth" in the Services Sector 204
"Productivity "in the Services Sector 205
"Productivity Growth" in the Services Sector 206
"New" Types of Services and Long-run Growth 206
Competition in the Services Sector 207
Global (International) Trade in the Services Sector 208

Chapter 9: Growth & Income Distribution 211

Functional Income Distribution 211
An Ideal Income Distribution 212
Pareto Optimum 212
Optimum Functional Income Distribution and Growth 213
Initial Case: Functional Income Distribution: with a "Given" Technology 215
1- "Efficiency Growth" & Income Distribution 216
2- "Innovation" & Income Distribution 217

Wage Rise in the Long-run & its Impact on Income Distribution 220
Technological Imperfections & Global Income Distribution 221
"Technology-intensive" & "Labor-intensive" Methods of Production 221
Concluding Remarks 222

Chapter 10: Growth or Development? 225

Growth 226
Development 227
Elements of the Development Theory 228
Development Economics 230

Chapter 11: Epilogue & a Suggestion 233

Developed Countries & Long-run Growth 234
LDCs & Long-run Growth 235
Producing New Technologies 235
More Efficient Use of Technologies 236
Efficient Use of Political, Institutional & Cultural Framework 236
A Suggestion to Boost Global Cempetition 237
Firms that "Only Produce Technology" 238
Suggestions for "Fairer" Income Distribution &
Increased Democracy at Work 240
Is "Unlimited Growth" Desirable? 241

Bibliography 243

While the countries that make up 80 percent of the world's population live in relative or absolute poverty, it is not possible to have Global peace and comfort.

H. Gürak

Introduction

Let's go back in time 60 years. Few homes had a TV-set. There were few TV-channels. Transmission was in black-and-white. Nowadays, we are hard put to find any home; even in the less developed countries that hasn't got a TV-set. Now we have multi-channel, "smart", color T.V.s. Communication systems have mushroomed. Sixty years ago you were lucky to have a "land-line". Today, almost everyone has a "mobile", if not a "smart phone". High quality communication through the internet is a fact of daily life. Automobiles, airplanes and household appliances are all becoming "smart". As we write, there is a "totally computerized" car being presented at the Istanbul motor show that doesn't need a driver!

Today's consumers enjoy the use of an immense variety of products that are of an unprecedented quality. "New" products are continuously being developed and introduced into the markets. In the developed countries, an increasingly large part of a person's income is being spent on unessential items. Items that we "want" but don't "need"! This signifies an increased per capita or nationwide prosperity.

The critical questions are: *Why do countries grow? Why do they grow at different rates? What is the role of "labor" or "human capital" or "capital" or "technological progress" in this process? Why is the income gap between the relatively richer and relatively poorer countries not converging? How could the development gap between countries converge?*

Until Solow's "*re-discovery*" of the vital role played by technological progress in growth in the 1950s, emphasis was placed on the "saving-investment" approach. Technology was considered as a "given". The only factor influencing long-run growth was an "exogenous growth in population". However and here's the rub, even in the case of a population growth, the markets were bound to saturate, sooner or later. As the markets saturate, the demand for products declines which would lead to a decline in profits. In this case the market would come to a static equilibrium and growth would end. But, in fact, this doesn't happen.

This approach in the pre-Solow period inevitably led to an undermining of technological progress' role in the growth process. Consequently, any analysis about the origin of technology and its contribution to growth was ignored for a long time. Yet the long-term profit rate shows no tendency to decline towards zero nor does the market show a tendency towards a "static" equilibrium. The economy is actually "dynamic" and "growing". Soon after Solow's "re-discovery", another significant contribution to growth theory was the "re-discovery" of the role of "the quality of laborer" or "human capital".

According to Mankiw (1995), the question; "Why do countries grow at different rates?" has a priority over the question "Why does growth occur?" Mankiw's view is misguiding. Unless we understand the internal dynamics of the growth process, we cannot understand why countries grow at different rates". An understanding of, where new technology is being developed, how the global technology markets function and how monopolistic control functions, is important in understanding the technology market "imperfections" (defects) in the global economy. If we don't correctly understand these issues, our knowledge of how economies function will be both inadequate and fruitless.

The Purpose & Scope of this Study

The driving force behind the book was the inadequacy of the available theories to explain the "growth process" in a satisfactory manner. The theories on growth, especially the neoclassical ones, fell far short in explaining and understanding real events. The assumptions of these so called "scientific" theories were unrealistic and "utopian". A Nobel Laureate admitted frankly that his "mechanical growth model" referred, in fact, to utopian economic relationships. The mainstream "scientific" theories were lacking in any awareness of historical developments. They did not take into account the impact of human values or the institutional setting on the economy.

Many attempts have been made to introduce new and more realistic theories about growth. However, there still seems to be scope for another perspective on the subject that would attempt to explain the growth and development processes in both the developed and less developed countries. We need new theories capable of explaining the growth process in a realistic manner which are able to account for the long run growth based on the new ideas (technologies) created by the creative mental labor.

The purpose of this study is to demonstrate that the "*cause(s)*" of long-run growth is "*technological progress*" which is the product of "*creative*" *mental labor*. The higher the "*quality of labor*" (*human capital*) *becomes*, the higher the potential to introduce new technology, given an appropriate cultural, institutional and technological infrastructure. In the absence of technological progress, long-run growth would inevitably end, while the rate of profit would continuously decline towards zero, cet. par.

Yet, the "real" evidence shows that growth is a continuous process. That is because; "*creative mental labor*" has an unlimited ability to introduce new technologies which secure the introduction of "*new products*" *and/or* "*new methods of production*". This is exactly the opposite of what the pessimist economists once suggested.

The creation of new technologies alone is not sufficient to secure welfare growth. There is also a need for an appropriately qualified labor force in order to make efficient use of the new technologies. Access to a qualified labor force is as important as the creation of a new technology because in the absence of an appropriately qualified labor force new technologies cannot be used efficiently.

In this work, Chapter-1 introduces some general information on growth. In Chapter-2, the key concepts of growth such as "productivity", "productivity increase e.g., "growth", the measurement of productivity will be presented and explained. These key concepts are critically important in understanding the "new" ideas and the alternative approach which will be presented later in our analysis. So the reader is asked to bear them in mind.

We will outline the "Classical" theories of growth by A. Smith, Ricardo, and Marx in Chapter-3. Up to date theories in regard to growth will be dealt with Chapter-4. It is interesting that practically all Classical economists were very aware of technological progress and its impact on growth together with the importance of the so called human capital. However, somehow since the 1870s, these key factors were ignored by economists, except for some like J. Schumpeter. Solow "re-discovered" them in the 1950s and introduced them into the growth model; but merely as an "external" factor; the source of which was "unknown". What a marvelous achievement in the name of "science"!

Neoclassical models shall be studied in two periods; the pre Solow and after. In the pre Solow era, growth models lacked any technological progress but had an extremely attractive feature; "equilibrium" which never had, or ever has had, anything to do with reality. In the post Solow era, the equilibrium "parable" and its fictitious assumptions survived at the expense of reality. But now technological progress had become an indispensable part of growth models. Nevertheless, the mainstream growth models continue to be largely a failure which is why we still need some "new" ideas.

Part-5 presents and discusses an "alternative growth model", followed by a short-run growth analysis in Chapter-6 and a long-run growth analysis in Chapter-7. Short-run growth can be realized with a "given" technology through a process called "efficiency increase". For long-run growth, on the other hand, technological progress which is the result of mental creativity is essential. In the absence of technological progress, no long-run growth can take place. The economy would saturate sooner or later, the profit rate would tend to fall towards zero and eventually growth would stop. This is not what happens in the real world because the creative mental labor constantly introduces "new" ideas and there seems to be no limit to the creative capabilities of the human mind.

A new analysis of "Growth in the Service Sector" in Chapter -8 is a unique part of the book as most growth analyses refer to the industrial sector. Normally, traditional growth models do not make a clear distinction between the industrial sector and service sector.

How technological progress and growth effects the functional and global income distribution will be the subject of Chapter-9.

In the final Chapter-10, we shall present a proposal for improvement in the global growth and development of nations, which is expected to help the convergence of the global income gap. Increased global competition, as suggested, would benefit all countries and enterprises.

Which Criteria Can We Use as a Comparison?

What criteria can we use to define a more useful model of growth? How can we determine which one of the growth models is more useful for a researcher or a politician or an economic decision-maker? How can we evaluate their usefulness?

So called "scientific" theories or models have to be, in principle;

1- Logical;
2- Consistent;
3- Explanatory and
4- Testable.

If the existing "older" theories/models of growth were capable in guiding us in explaining the growth process or in comprehending the difference between the different growth rates or in formulating policies to achieve higher growth rates in a satisfactorily, there would be no incentive for us to research alternative models. Yet, as we all know, the existing "older" and more recent theories of growth fall short in explaining many of these aspects. These will be discussed in Chapter-4 and Chapter-5.

How can we know if the "new" ideas in this book are better alternatives? To find out, we can make an assessment of the alternative approaches. Generally the most suitable method seems to be to compare the two hypothesis using the statistical "*likelihood principle*" and see which one is closer to reality. The one which is simpler to explain, has less fictitious assumptions and is more representative of the actual economic relationships should be considered as "evidentially favorable". Thus this theory/model should be considered "more useful". For example; if a theory assumes "homogenous goods" while the alternative assumes "multiple goods", the latter is preferable as more realistic thus it is a better theory or model.

If one theory assumes the role of having "perfect knowledge", while the alternative assumes knowledge is "imperfect", the latter should be considered as a more practical theory as it is more relevant to the "real situation". A theory assuming that technological progress is an "endogenous" factor should get more credits than the one assuming it to be an "exogenous" factor. This way all the pros and cons about different theories or models can be estimated and the theories can be classified according to their credits gained on the "usefulness" criterion.

Based on the "usefulness" criterion, as stated above, we claim that the "new" ideas throughout this book are "more useful" than the "older" theories or models, and are capable of giving a better account of real economic relationships. To prove that, we do not require sophisticated mathematical or econometric models. All we have to do is to ask the "actual" economic actors in the markets e.g. the business-people and managers who make the economic decisions that affect us all and have to account for the consequences of their decisions. After all we are talking about economic relationships which affect all of us.

Briefly, the "new" ideas presented in this book are:

1. Logically consistent and explanatory.
2. Assumptions are more realistic.
3. Closer to "actual" economic relationships.
4. Allowing logical, consistent and realistic predictions.

Just for the record; the author of this book is in no way claiming to be superior in any way to any other economist in terms of his understanding or his natural talents with regard to economic matters. On the contrary, many economists have already proved by their contributions that they are smarter and more creative than the author in many ways. However, a large percentage of them have the disadvantage of not having any practical experience in actual economic relationships. They, in general, learn their economic knowledge from text-books and then create imaginary relationships flavored heavily by the Anglo-Saxon ideology. In other words, because of their lack of ability to think *experientially* they seem to be at the starting gate.

The main difference between the author and them is the significant advantage he has in having 20 years of experience in the real economic sector, especially in the Global business network. This experience is invaluable and cannot be learned from text-books. If a person considers that text-book learning is sufficient in the mastering of any branch of science, in the making of analyses and predictions, then that person is making a major mistake and should not be allowed to make decisions that affect the lives of millions of people.

Let's consider Solow's contribution: what is ordinary information for the practicing business people, i.e. the role of technological progress in growth and competition, was only "re-discovered" by him in the 1950s. Moreover, Solow's model could not give us a hint on the origin of technological progress. It was assumed to drop into the economy as "manna from heaven". Yet, for experienced people, technological progress has always been there; as an "endogenous" not an exogenous factor, as he suggested. Nevertheless, the textbook economists of the same ilk did not hesitate to award him the Nobel Prize.

The principal *"non-mainstream"* and *"non-book-learned"* messages can be summarized as:

1. Technologies are produced by "creatively" QUALIFIED LABOURERS.
2. Technologies are used by QUALIFIED LABOURERS.
3. Long-run growth cannot be sustained without QUALIFIED LABOURERS.

Chapter-1
Growth Process Worldwide

Historically, we see that countries perform at different *growth rates*. Two hundred years ago the gap between development and income amongst countries was not as huge as it is today (see Figure: 1-1). The estimate of GDP per capita in 1700 showed that per capita income everywhere the world was very similar. There were small differences between the US, China and India. From 1700 until 1820, almost no change in world GDP per capita occurred. (See Vasquez;2003;90).

Figure: 1-1 World per capita GDP*

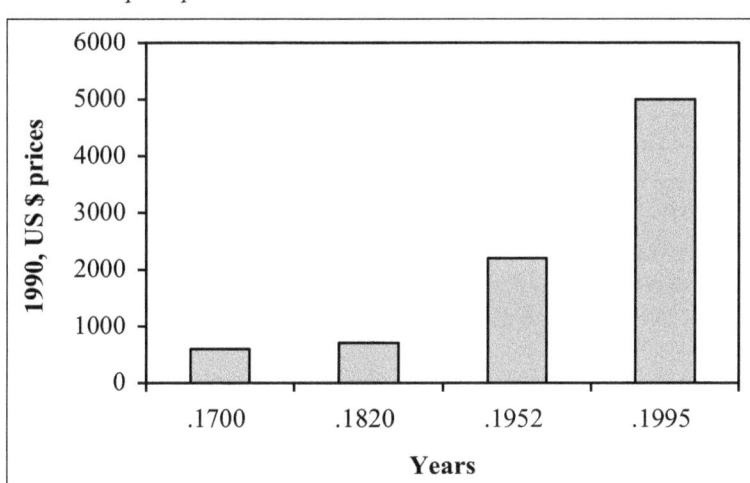

Source: A. Maddison, Monitoring the World Economy: 1820-1992 *in* Vasquez, I.0 (2003), Kapitalizm ve Küresel Refah, p. 90, Figure: 5.1. * quoted from the Turkish version of the book.

During the 1700s, differences in productivity growth began to occur; compared to China, Japan and Russia, per capita income in Europe and the US almost doubled. The economic boom of the 19th century tripled the living standards in Europe and quadrupled them in the USA (Vasquez;2003;90).

Table: 1.1 shows us the per capita income in the 1700s and 1800s. The same Table also shows that the income gap in the 1700s and 1800s was much smaller than it is today on the basis of 1960 US Dollar prices. Towards the end of the 1800s per capita income was around $ 190-230 in Sweden, $ 160-200 in Japan, while it was $ 170-210 in Egypt, $ 130-160 in India, $ 240-280 in Jamaica in 1832.

Table: 1.1 The per capita GDP estimates in some countries in the pre-industrial period. (in US $ in 1960 prices)

Developed countries	Period	Per capita GDP
England	1700	160-200
USA	1710	200-260
France	1781-90	170-200
Russia	1860	160-200
Sweden	1860	190-230
Japan	1885	160-200
Developing countries		
Egypt	1887	170-210
Ghana	1891	90-150
India	1900	130-160
Iran	1900	140-220
Jamaica	1832	240-280
Mexico	1900	150-190
Philippines	1902	170-210

Source: P. Bairoch; in C. Freeman-L. Soete (2003) Yenilik İktisadı, p. 364, Table: 13.1

Table 1.2 shows how the per capita income gap widened from 1750 to 1977. While the ratio between the richest and the poorest was 1.8 in 1977, this ratio increased to 29.1 in 127 years; a small difference turned into a yawning chasm.

Table: 1.2 The development of the per capita GNP from 1750 to 1977

Years	Developed	Countries	Third	World		Deficit
	(1) Total (billion $)	(2) Per Capita	(3) Total (billion $)	(4) Per Capita	(5) = (2)/(4)	(6) The ratio of the richest to the poorest
1750	35	182	112	188	1.0	1.8
1800	47	198	137	188	1.1	1.8
1830	67	237	150	183	1.3	2.8
1860	118	324	159	174	1.9	4.5
1913	430	662	217	192	3.4	10.4
1950	889	1054	335	203	5.2	17.9
1960	1394	1453	514	250	5.8	20.0
1970	2386	2229	800	380	7.2	25.7
1977	2108	2737	1082	355	7.7	29.1

Source: P. Bairoch (1981); quoted from G. Dosi, et. al. (1992); in C. Freeman-L. Soete (2003) Yenilik İktisadı, p. 365, Table: 13.2

> In Latin America, only three countries have grown faster during the 1990s than in the 1950-80 period. One of these three was Argentina, a country whose hopes of economic salvation through financial integration with the world economy now lie in ruins. A second was Uruguay, also in deep trouble. Only Chile looks like a long-term success. Among the former socialist economies, real output still stands below 1990 levels in all but four of them. And poverty rates remain higher than in 1990 even in Poland, unquestionably the most successful of the East European countries. In sub-Saharan Africa, results remain very disappointing, and far worse than those obtained prior to the late 1970s.
>
> D. Rodrik, 2002, After Neoliberalism, What?
http://ksghome.harvard.edu/~drodrik/after%20neoliberalism.pdf

Table: 1.3 demonstrates the substantial changes in the level of income and the qualitative structure of the US society as a result of 25 extra-ordinary developments in the 20th century.

Due to "new" technologies, the productivity per laborer and societal wealth has been continuously growing. In spite of many serious environmental problems, nature still provides us with various raw materials which qualified laborers can turn into new products and new production methods. As long as the creative ability of human brain and natural resources do not exhaust, growth and development will continue.

Excluding the hyped-up success stories of some newly industrialized countries in the Middle Income group, developed countries seem to be the major beneficiaries of the recent improvements in Global income. On the other hand, the position of the less developed countries seems to have worsened in terms of global income distribution. Nowadays, the rich countries are richer and the poor ones poorer." (See. Table: 1.4).

The worldwide growth prospect does not seem bright. According to a UNCTAD Report:

> *"The world economy has seen a modest improvement in growth in 2014, although it will remain significantly below its pre-crisis highs. Its growth rate of around 2.3 per cent in 2012 and 2013 is projected to rise to 2.5-3 per cent in 2014. This mild increase is essentially due to growth in developed countries accelerating from 1.3 per cent in 2013 to around 1.8 per cent in 2014. Developing countries as a whole are likely to repeat their performance of the previous years, growing at between 4.5 and 5 per cent, while in the transition economies growth is forecast to further decelerate to around 1 per cent, from an already weak performance in 2013."* (UNCTAD; 2014; Trade and Development Report; 1). (see Table: 1.5).

Table 1.3 25 breakthrough developments in the US in 20th century.

	1900-20[a]	1995-98[b]
Life expectancy (years)	47	77
Infant death (per 1000 birth)	100	7
Death from contagious diseases (per 100,000 persons)	700	50
Heart diseases (per 100,000 adjusted to age caused deaths)	307 (1950)	126
Per capita GDP (in $, 1998)	4,800 $	31,500 $
Wages in manufacturing (in $, 1998)	3.40 $	12.50 $
Household assets (in billion $, 1998)	6 $ (1945)	41 $
Poverty rate (percent of USA households)	40	13
Weekly working hours	50	35
Agricultural workers (percent of total labor-force)	35	2.5
TV-set owners (percent of USA households)	0	98
House owners (percent of USA households)	46	66
Electrification (percent of USA households)	8	99
Telephone calls (per capita calls per annum)	40	2,300
Transport vehicles (percent of USA households)	1	91
Patents granted	25,000	150,000
High-school graduation (percent of adults)	22	88
Deaths caused by accidents (per 100,000 persons)	88	34
Wheat price (per kilo for hours employed)	4.1	0.2
Grades of women graduate degrees (percent of grades)	34	55
Blacks' income (per capita, in 1997 $ prices)	1,200 $	12,400 $
Total population registered in the USA (in millions)	76	265
Air pollution (microgram lead per 100 cubic meter air)	135 (1977)	4
Speed of computer (millions directives per second)	0.02 (1976)	700
Computer ownership (percent of USA households)	1 (1980)	44
a: Figures are based on the oldest years available. b: Figures are based on the latest years available.		

Source: Vasquez, I.0 (2003), Kapitalizm ve Küresel Refah, p. 62, Table: 4.1.
Note: Table is translated from Turkish version by the author.

Table: 1.4 Development of Per Capita Income in World 1970-2000

	1970	1980	2000
Per capita income (in 2000; US $)			
Developed countries	11,001	16,323	26,843
Developing countries	884	936	1,162
Sub-Saharan Africa	757	675	493
The least developed	410	366	306
Relative per capita income (%)			
Developed countries	100.0	100.0	100.0
Developing countries	7.0	5.0	3.9
Sub-Saharan Africa	6.0	3.6	1.6
The least developed	3.2	1.9	1.0
High-debt countries	4.4	2.5	1.0

Source: F. Beaugrand, 2004, "And Schumpeter Said, "This Is How Thou Shalt Grow", IMF Working Paper, March. in: E.Yeldan, et. al.; 2012; 34; Table: 1

Table: 1.5 World Output Growth, 2006-2014 (Annual percentage change)

Region	2006	2007	2008	2009	2010	2011	2012	2013	2014[a]
World	4.1	4.0	1.5	-2.1	4.1	2.8	2.3	2.3	2.7
Developed Countries	2.8	2.5	0.0	-3.7	2.6	1.4	1.1	1.3	1.8
Developing Countries	7.7	8.0	5.4	2.6	7.8	6.0	4.7	4.6	4.7

Source: UNCTAD, 2014, Trade and Development Report, Table: 1.1
a: Forecast

All About Growth

Let's recap some basic issues. At the micro-economic level, growth for a firm (or an enterprise) may imply an increase in the annual rate of profit, the size of the profit, the market-value of the firm, an increase in the quantity of products sold, increased plant-capacity utilization, or some other criterion, in compliance with the interests of owners. A firm's target is to make profits. A firm, by its nature, does not aim to increase employment, or reduce the trade deficit or to achieve growth at the national economy level.

Every firm is a legally registered entity in "a" country in which it is liable to that country's rules, regulations and laws. The firms together comprise the national-economy which is the subject of macro-economics. For example, the employment issue, balance-of-payments, exports, etc., are all macro-economic issues.

Our concern throughout this book is, basically, to study and analyze a single "macro-economic issue, the growth of the national economy as a whole. However, we shall closely study what happens at the micro-level, or firm-level, since the performance of the national economy is dependent upon these firms.

What does growth at macro-economic level mean? How does it come about? Which forces drive growth, especially in the long-run? Which are the forces hampering global growth? These are the major questions we have to deal with.

Let's start by taking a look at the definition of growth in the available literature. According to I. Parasiz growth implies an increase in the potential full-employment output in time. (Parasız;1996;223). For Hatipoglu growth means an increase in the national income of the developed countries. (Hatipoğlu;1993;36). Üstünel describes growth as an increase in the productive capacity of a country. (Üstünel;1990;218). For Lipsey growth implies long-run changes in the per capita factor productivity. (Lipsey, et.al;1990;333). According to Manisalı, growth means an annual increase in the Gross Domestic Product (GDP). (Manisalı;1994;164).

Throughout this book, growth at a national economy level will be considered as *the change in added value (VA[1]), per worker and per unit of time*. The change may be positive or negative depending on the direction of the change of the added-value. This process of growth will be studied under two main sub-headings; "short-run" and "long-run/". Our main objective is the study of the/"long-term/ growth" of the national economy.

How does growth occur in a national economy?

To achieve a positive growth rate, the amount of the "added-value" has to increase. "*Over the last few decades, the list of proposed panaceas for growth in per-capita income included*:

1- *High rates of physical capital investment,*
2- *Rapid human capital accumulation,*
3- *Low income inequality,*
4- *Low fertility,*
5- *Being located far from equator,*
6- *A low incidence of tropical diseases,*
7- *Access to the see,*
8- *Favorable weather pattern,*

---

[1] Added-value (VA) = Gross wages + Gross profits.

9- *Hands-off governments,*
10- *Trade-policy openness,*
11- *Capital-market development,*
12- *Political freedom,*
13- *Economic freedom,*
14- *Ethnic homogeneity,*
15- *British colonial origins,*
16- *A common-law legal system,*
17- *The protection of property rights and the rule of law,*
18- *Good governance,*
19- *Political stability,*
20- *Infrastructure,*
21- *Market determined prices (including exchange rates),*
22- *Foreign direct investment,*
23- *Suitably conditioned foreign aid."* (Wacziarg;2002;907).

Or, perhaps, the global economic order as it works?

All these panaceas for growth were developed and passionately proposed by the Western minded experts and have been practiced wholly or to some extent in developing countries. What was the outcome? A success story? Unfortunately not.

Who is (are) liable for the failure? Could it be the panaceas themselves on the grounds that the inadequacy of the propositions was incapable of providing solid and permanent remedies? Or, perhaps, it was the manner in which the global economy works?

"Savings = Investment" Paradox

Some renowned contemporary economists such as Barro, Sala-i-Martin and Levine, believe that the source of growth are "exogenous" technological changes as Solow suggested. The "Total Factor Productivity" (TFP) criterion is considered as the correct way to estimate the growth rate. P. Romer promotes an "endogenous" new knowledge creation process as the source of growth. According to Skousen, savings imply the production of investment goods[2]. We shall study all these views later in the book. But, for now, let's take a closer look at a phenomenon, a "legendary assumption", inherited from the Classical economists.

---

2 *"... there are two kinds of production – production of consumer goods (consumption) and production of investment goods (saving)."* (Skousen;2001;57).

Once upon a time, in the time of Classical economists, it was assumed that only the capitalists saved and they invested all their profits to make more profits. That was considered a "rational" form of behavior because money was considered as a "medium of exchange", only. Why keep the money idle under these circumstances?

As A. Smith suggested, the amount of investment was determined by the amount of savings. In order to obtain a greater income, the capitalist should invest his/her savings and acquire more profits (Adelman;1972;34-35). So investment of all profits i.e. savings was a "rational" behavior and thereby the well-known S=I equation was introduced into economic theory. For the Classical economists in general, money was simply a means of exchange; it wasn't a means to store value; nor a source of economic-political power; nor a precautionary factor. Therefore, a rational capitalist had only one choice; to invest. More savings; thus more profits required more investments. And more investments would imply growth. So only the capitalists saved.

What is the probability of the S=I equation assumed by "Classical economists" coming true today? Do investments really automatically rise as the savings increase? Or vice versa; do investments actually fall as the savings decline? Or should the basic question simply be; "what causes investments to rise or to decline?"

Over time, economic thinking and the comprehension of economic relationships have undergone some radical changes; but the assumed S=I equation has managed to survive. Contemporary text-books on macroeconomics and growth theory, promote the "savings equals investment" paradigm and many politicians, technocrats and bureaucrats, i.e. the country's decision makers, subscribe to this view. If the saving problem is solved, they assume then the growth process will continue.

According to contemporary understanding total savings in a country equals the sum of total public and private savings. In other words, total savings (S) is equal to a nation's income (Y), minus consumption (C) and government expenditures (G), i.e. S=Y-C-G=I. In other words, what is not spent on consumption and by the state make up the total savings and this portion goes to investment (I). Though it seems like a logical conclusion at first sight, neither the Classical S=I assumption nor the contemporary S=I assumption reflect fully actual economics as we practice it. There are many reasons for that and here are some of them:

1- In contrast to the assumption of the Classical economists, not only the capitalists but also "others", including the wage-earners, do actually save.
2- All savings do not have to be invested. Mainly because money is not simply a medium of exchange. It is also a medium of "accumulation and the source of economic-political power, regardless whether it is employed in the supply of

products or not. In addition, money can be saved (kept idle) for precautionary purposes.
3- Investment funds can be created within the monetary system through a Central Bank or other banks or financial institutions.
4- Whenever domestic funds are inadequate, foreign financial funds in other countries can be utilized. In fact, sometimes foreign funds may be cheaper than the domestic funds.
5- Increased savings at the national level, cet. par. implies less expenditure on consumption which in turn implies a reduced amount of profit on average. When profits decline, there would be less incentive to invest which in turn would adversely affect growth. At the Global level a small nation may benefit from increased savings. But if many countries attempt to follow this process, the Global economy as a whole may be affected adversely.

Keynes did not share the view of the Classical economists that savings equal investments. He considered that there was no guarantee that all savings would be invested. However, Skousen criticizes Keynes and says:

"Historically, the evidence is overwhelming: higher saving rates lead to higher growth rates, just the opposite of the standard Keynesian prediction." (Skousen;2001;367).

According to Skousen:

"The problem is that Keynesians treat savings as if it disappears from the economy, that it is simply hoarded or left languishing in bank vaults. In reality, saving is simply another form of spending, not on current consumption, but on future consumption." (Skousen;2001;368).

According to Keynes' opponents, more savings are expected to reduce the interest rate which is expected to encourage more investments. But, experience shows us that this doesn't necessarily and automatically occur. In other words, lower interest rate does not automatically lead to more investment. Today, the interest rates in the USA, the EU and Japan are at quite low levels; in fact, frequently, we observe negative interest rates. Yet, the USA, the EU and the Japanese economies do not seem to grow as expected. Why is that?

For those economists who think like Skousen, when savings increases and the interest rate fall, technological innovations keep pace and increase. This is a serious mistake. The lower interest rate may encourage more research due to a reduced cost of financing the research and development of new products. But there is no guarantee that these innovations will be economically successful in sustaining long-run growth. Is the low growth rate problem in the USA, the EU or Japan a problem of "high interest rate? Certainly not!

In fact, the interest rate is not a determinant factor in technological progress. The source of technological progress which is the major determinant for long-run growth is a product of the "creative mental labor". In the absence of "creative mental labor" there would be no innovations at all. In other words, the origin of the long-run sustained growth is the creative capability of the human mind endowed with skills, talents and education which are referred to, mistakenly, by some economists, as "human capital".

In a competitive environment, firms strive to gain advantage over their competitors by introducing new technologies. A new technology may be a cost-saving new process or the introduction of "new" products. The degree of economic success of these innovations is another story which depends on the abilities of the managers of these innovations.

In summary, we can simply claim that increased savings either may or may not increase investments. The decline in the interest rate may or may not trigger an increase in investments.

Associating a rise in growth rate to increased savings seems to be, to say the least, an over-optimistic view. The impacts of factors such as the quality of the labor force, the rate of technological change, effective increase in demand for products, the prices and qualities of the competing products, the global interests of the giant and globally dominant firms should be evaluated carefully.

On Added-value Criterion

As previously mentioned, the growth of a national economy will be considered in terms of the total added-value (VA) consisting of gross profits (π) and gross wages (w), e.g. $VA=w*L + \pi^3$. The reason for the use of the added-value criterion is the measurement of incomes in a "similar" way in terms of Gross Domestic Product (GDP) within a specific time period.

"Similar" because the components of the GDP may vary. For example, according to the so called expenditure method, GDP consists of the total market value of all goods and services produced in a given time-period, say a year, in a country.

$$GDP = C + I + G + (X - M)$$

In terms of values, C, symbolizes gross consumption; I, gross investment; G, government spending; X; exports and M, imports.

Another way of measuring GDP is to measure total income. *"The US "National Income and Expenditure Accounts" divide incomes into five categories:*

---

3   No depreciaiton or taxes unless otherwise stated specifically.

1. *Wages, salaries, and supplementary labor income*
2. *Corporate profits*
3. *Interest and miscellaneous investment income*
4. *Farmers' income*
5. *Income from non-farm unincorporated businesses*

These five income components sum to net domestic income at factor cost." (http://en.wikipedia.org/wiki/Gross_domestic_product\) 2013-12-09.

Total factor income may also be estimated as:

$GDP = W + \pi + PI + R + i$

W, symbolizes wages; π, profits; PI, proprietor's income; R, rental income and i, interest.

Another way of measuring the economic activity in a country is to measure the *Gross Domestic Income (GDI)* which consists of all wages, profits and taxes minus transfers/subsidies.

$GDI = W + \pi + t - t/s$

As mentioned before, we shall use the criterion of changes in the added-value to measure growth.

$VA = GDP = w*L + \pi$

The Relationship Between the Laborer – Technological Progress & Growth

We can look at labor in two ways;

1- as *"intellectual or /mental"* labor; or
2- as *" simple physical"* labor.

Physical labor has a limited contribution to the growth process or to adding value. Simple physical labor is merely a response to orders from the brain. It is the fount of our intellectual and creative endeavors. No-one knows exactly how this happens but it certainly does happen! This is what we are defining as "intellectual or mental labor". When we consider the result of this aspect of the brain's activity in regard to economics, we can say that the brain produces *"productive knowledge"* and this happens continuously. Another way of defining this "productive knowledge" is simply *"technology"*. It may be stating the obvious but I think everyone would agree that the real and continuous source of economic growth, in the real world, is "technology". If "mental labor" had not led to a continuous technological progress over the past millennia, we would be fireless, wheel-less, car-less and any other 'less' you want to add. In fact we probably would not be around, to worry about a "realistic" growth theory!

So when we ask about what causes long-run growth, given the natural resources and the physical labor, we have to say "technology". "Technological progress" is the product of the human creative intellect (Gürak;2000-a).

What empirical evidence have we got for this? Look around you. "New" products (goods and services) and new production methods, e.g., new technologies, are appearing at an unprecedented rate. If there had not been any technological progress; if only "homogenous goods" were being produced; markets would have saturated, profit rates would have decreased and new investment would have stopped. Contrary to what the pessimists say, the reason why production has grown, a continuous growth process is occurring and average profit rates do not tend to decrease to zero in the long-run, are all due to "technological advances". As stated by Toffler (1992; 96), *neither added value nor wealth can be created in the economy without intellectual labor.* Drucker also states in one of his works (1995; 30) that the origin of wealth is a "humane thing"; this humane thing is KNOWLEDGE" (author's emphasis).

Technological innovation is generally produced by the Developed Country Firms (DCFs). This is because the developed countries possess more highly qualified human resources and a more appropriate infrastructure. Consequently more importance is placed upon and greater resources are allocated to R&D. Table: 1-6 shows the amount spent on R&D per capita according to purchasing power parity/(PPP) in some countries.

Table: 1-6 The amount spent for the R&D as per capita in some countries (according to purchasing power parity, US$)

	2000	2001	2002	2003
EU-25 countries	398.4	423.7	443.3	–
Japan	775.5	816.5	838.5	–
USA	939.0	962.9	961.4	977.7
Turkey	43.1	43.8	43.0	–
Mexico	33.6	35.8	–	–
South Korea	395.8	452.1	477.2	518.9

Source: OECD, Main Science and Technology Indicators, 2004-a; p. 19, quoted from data on the Table: 04

According to the data supplied by the OECD, the EU and South Korea spend about 10 times more per capita on R&D than middle income countries such as

Turkey and Mexico. In Japan and the US, the amount per capita spent for R&D is much higher. As a result, there is more new technology and patent acquisition in the Developed Countries (DCs). (see Table: 1-7).

Table: 1-7 The number of patent applications by country (2002)

EU-25 countries	60,698
EU- 15 countries	60,158
Japan	21,248
USA	44,427
Turkey	70

Source: EU Science and Technology Indicators, 2005 Edition; data quoted from Table: 7-1 on p. 100 and 101.

Thanks to the patent right granted for a technological innovation, the patent holder secures a monopoly and a temporary competitive advantage in the market. He/she also acquires a profit rate above the market average. The DCFs are way ahead in terms of technological innovation, especially in the high technology and highly qualified labor sectors. If we take a look at the patent applications related to bio-technology; there is a marked gap between firms in the DCs and everyone else (see Table: 1-8).

Table: 1-8 The number of patent applications in the bio-technology industry by country

	1999	2000	2001
EU-25 countries	2,025	2,102	1,995
Japan	588	705	711
USA	2,886	2,822	2,303
Turkey	0	1	1
Mexico	4	2	2
Russian Federation	15	15	16
China	11	20	35
Argentina	3	4	3
Rumania	0	0	0
Slovenia	0	2	2

Source: OECD, Main Science and Technology Indicators, 2004-a; data quoted from Table: 68 on p. 52 and 62.

There is a direct relationship between economic growth, mental labor and technological innovation in the long-run. Obviously other factors have an impact on the growth process both in the long and short term. We will discuss these as they become relevant.

What Role Does Investment Play in Regard to Technology and Long-run Growth?

Considering technology and investment as independent factors is bound to lead to problems. In the absence of new technologies, the growth process would stop in the long-run; simply because the new investment opportunities which lead to new and increased profit opportunities would become exhausted. This, a mainstream economist may say, is the "equilibrium" state of economy. However, if laborer is creative and consumers continue their unbridled passion for "consuming", obviously "new" products and "new" production methods will continue to emerge. Entrepreneurs do not want to be wiped out. They want to stay in a permanently competitive market. They want to avail themselves of, and profit from, "new" opportunities and look for new "investments". So, briefly put, new technologies lead to long-run investment.

To sum up;

1. The source of long-run growth is technological progress.
2. The source of technological progress is "qualified labor".

Access to and the Use of Technology

In mainstream models technology is generally accepted as a "given". It is assumed that everybody can access this technology without restriction. Some modern textbooks still stand by this assumption i.e. that anyone can have unrestricted access to "perfect production information" is unfortunately, not the case.

In reality technology is not freely available to everyone and cannot be used "without restriction". In fact, there are many serious restrictions on accessing new technology. P. Romer (1990), suggests the use of a technology by someone else, does not decrease the "amount of technology" available to others. For him, technology is a "non-rival" good which can be used by anyone, anytime. Once a new "technology" is produced, the marginal cost of the re-using it is zero. But, he also goes on to say access to the technology can be partially excludable by patent laws. Isn't this a bit like saying "You can have as much of the pie as you like, BUT only if I have the knife and I cut it and I choose the bit you can have.

One has to agree with Romer's claim that technology is a "non-rival good". After all, technology is a body of knowledge related to production. Sharing this knowledge with others would enhance the font of cumulative knowledge. Of course knowledge can be more enhanced by sharing it. It is likely, all things being equal, that in two communities with the same educational and infrastructure level, the one with a thousand members would contribute more to the knowledge base of a society than the one with only a hundred. The greater the knowledge is available to a great number of users would clearly benefit the global economy. But the right to access and use that knowledge (technology) is restricted by patent laws. These patent restrictions simply serve commercial interests; that is because the owner of the patent is protected by law for a certain period time and will have a monopoly in the market.

After a new technology comes on the scene, two things can happen. Some competitors will develop a similar but slightly different "new" technology, in order to compete against the monopoly; or they purchase or lease the right to the patented technology. A patent holding "monopolist" would be reluctant about the latter option, since he/she would not want to lose their advantage. So, inevitably the competing firms develop a similar "new" technology. Otherwise, they risk being unable to compete in the open market. For sure, they will have to allocate resources for this "new re-invention". Ultimately some will succeed, while others fail.

Who has the potential to compete in the open market by producing "new" technology?

The producers with greatest potential are easy to identify; generally it's the companies in the DCs with global operations. These DCFs have, normally, access to both "qualified labor" and an advanced scientific, technological/ infrastructure. In addition, their governments are better organized and more efficient; and their cultural framework similar to one another. They can also produce a similar "new" technology within a relatively short period of time. This production of similar new technology is, in a way, a waste of resources. The desired technology is already in place. Employing resources for the development of something that already exist is irrational. But, this is consequence of the internal dynamics of the capitalistic competitive system. It is likely in the DCs, that over time, the monopolistic market system will change to an oligopolistic one.

For the other 80 percent of the world's population, who live in the developing countries (LDCs), the issue it's a very different one. In these countries the qualified labor, the scientific and technological infrastructure, the cultural framework, or the governmental system, is not at a sufficient level in order to sustain their own technological development at the contemporary levels. Let alone inventing or developing a "new" technology, the one's they have are often not used efficiently. Some ordinary technology in the DCs is often considered high-level for producers in the LDCs. If the aforementioned factors weren't in play, the technological and economical gap between the developed and the developing countries would not be so wide.

Producers in the LDCs have to pay for technology under the present global economic conditions. They do this by paying royalties or license fees for the transfer of any technology that they don't possess. However this monetary payment is just the beginning. "Imperfections" (improper methods) in the technology market hold to ransom, not only producers in the LDCs but also their whole economy as they are dependent on globally operating firms. This process not only prevents "new" competitors from emerging in the LDCs but also prevents an increase in global competition. The reason is simply; *the global division of labor does not have the inherent features to enable a process of technology transfer which will increase global competition. The global economy is not only increasingly coming under the control of the developed country firms, but also the global economic order is being shaped in accordance with their global interests.*

Laborer, Knowledge & Growth

Labor inputs have been the basic resource in long-run economic growth and in the increase of wealth in communities for aeonian. For millennia the font of human knowledge has been growing. So too, has the income and wealth of communities. The rate of growth and the accumulation of knowledge accelerated during the Industrial Revolution. Today, with the advent of new communication technologies, both production and the availability of information and knowledge have accelerated. Consequently the need for more qualified laborers has grown rapidly. Therefore, more qualified laborer will enable contemporary technology to be used more efficiently. In other words, more qualified laborers mean increased productivity and an advanced country. Of course, this means that companies have to engage in a "lifetime" commitment to educate their "laborers".

World War-2 had physically destroyed both Germany and Japan. But, both countries have grown at a fast pace and are now ranked among the wealthiest economies in the world. They had the required productive knowledge and qualified labor force to rebuild the country and boost production. However, Turkey, who did not participate in World War II, neither was it physically destroyed, is still one of the lower income countries of Europe. The main reasons for this are:

1- A poor level of education.
2- A poorly endowed labor force. (i.e. a low level of qualified labor force).
3- An underdeveloped technological infrastructure.

If the amount of a country's qualified laborers is high enough, the technological development and the income of that country is highly likely to be high. Having remarked on this, it appears self-evident that there is close relationship between qualified labor force and the level of education, which in turn determine the level of development of a particular country.[4]

There are various ways to measure the qualification level of the labor force. We can use "total years spent in education" as a criterion, but the possibility of a sound result is low. It doesn't take into account; the quality of education, the quality of the teacher or the knowledge and skill set of the student. Furthermore, how can we measure *"experience"* properly? Experience is an important factor, in the advent of qualified laborer. "Experience" is a multifaceted concept influenced by; personal skills, abilities, career progress, education and years spent at job. Therefore, it is easier to use the term "qualified laborer" as an analytical concept, instead of trying to measure each element involved in experience mathematically or statistically.

Qualified Labor or "Human Capital"

What is meant by *"qualified labor"* is; the basic qualifications that the laborer gains by means of formal, informal education or training, enhanced by experience. The term *"qualified labor force"* refers to the "sum" of laborers in a community. So, qualified labor is the "service" provided by a qualified laborer.

In mainstream models there are "two production factors", capital (K) and labor (L). However a new production factor began to appear in the 1950s, "technological progress" (A). In the 1960s another production factor began to emerge; *"human capital"* (H) e.g. *"quality of laborer"*.

This, "H", gave rise to some confusion. Does the term, "human capital", apply to *"human"* or *"capital"*? This needs clarification. These two terms are opposing forces in the distribution of income. If we look at the term grammatically; it must be related to both. Is this possible in economic terms?

Without a doubt, "human capital" essentially refers to humans, or more precisely, the labor force. "Human capital" refers to the knowledge, skills and

---

4   Former Soviet Union and Eastern Europe countries are exceptions, because said countries experiences an transition period and they have neither corporate and cultural infrastructure nor "competitive" firms.

experiences of the laborer (qualified labor). In fact, we can use the term *"quality of labor"* instead of "human capital". It is exactly the same. Some economists take a different approach. In their analyses, they use these two terms (i.e. 'human capital' and 'labor') as completely different inputs in the production process.

In this work, *"human capital", "quality of labor"* and *"qualified labor"* means the same thing. For reasons of brevity we will use the term *"qualified labor"* where applicable. In economic terms, to tag "human" onto "capital" is irrational, to say the least.

So why the terms "human and capital" have been joined together? We think there are three reasons for this.

1- The first is ideological. Some economists might have thought that the use of the concept, "quality of labor" would serve the interests of Marxist ideology by emphasizing the importance of the laborer in the production process. Their choice of "capital" next to "human" may seem reasonable to them. This should come as no surprise. Economics has always been subject to an ideological hegemony. Many renowned economists are well known as passionate defenders of "utopian" economic models. As the "human capital" nomenclature agreed with their various standpoints (theory), it became a term that was accepted *without criticism* and subsequently became a revered and prominent phrase in the academic world of economics.

2- The second reason for the adoption of the new term "human capital" may simply be due to a form of "academic face saving". If the concept of the "quality of labor" was acknowledged as significant in explaining economic transactions, the entire spectrum of the existing economic theories had to be re-considered. With the use of the concept of "human capital", the existing theories and the models remained intact. So it seemed easier to inject the concept of "human capital" into the then prevailing theories as an additional production factor. As a result, the mainstream models remained viable. They had got rid of a potential threat. They were still in the game!

3- The third and final reason may be one of optimism. The importance of capital is paramount in production. Now one had to estimate which type of "capital" makes the higher contribution; "human capital" or "production capital". However, to do this, one has to be aware, that human capital is in fact a concept related to the "laborer" and one has to be able to demonstrate this without bias. It is necessary to show clearly and unquestionably that labor (L) and human capital (H) is one and the same thing. These two concepts originate from the laborer or "the worker". Many economists are not happy with this way of thinking on ideological grounds.

The notion of "human capital" and "qualified labor" are the "same"; as two sides of the same coin.

Creative Labor and Growth

The importance of qualified laborer in the efficient use of modern technology is huge. However, the mere existence of qualified laborer is not sufficient to secure *long-run* growth. After a while markets would become saturated and growth would stop. Long-run growth can only be sustained by introducing *new technologies* and in order to introduce a new technology, we need *creative laborers*. So, for long-run growth, the most important factor is *creative qualified laborers*.

Unqualified Labor

If we use *"qualified labor"* as a benchmark, then *"unqualified labor"* has to refer to an "uneducated", "unskilled" and "inexperienced" laborer. Such a definition is neither rational, nor realistic or scientific. In the strict sense of the meaning "unqualified" labor can only exists in theory. Today, everyone has some form of formal, informal education, experience or skill set. Everyone is "qualified" in some way. Realistically, in theoretical terms, we can only refer to the *"relative" differences in the quality of labor or laborer*. Therefore, we have to define "unqualified" labor in a different way. This definition must take into account the variation in the qualified labor. Instead of using the "qualified-unqualified" dichotomy, maybe "less" or "more"- qualified would be appropriate. Simply using the terms "laborer (L)" and "human capital (H)" would lead to misleading analyses and unproductive conclusions.

Technological Progress & the Global Economy

In general terms, we can divide firms into two groups.

1. *The creators and owners of technology.*
2. *The "users" of available technology.*

Firms capable of creating new technology are, generally, based in the developed countries (DCs). The reasons for this is simple, namely; the availability of qualified laborers, an appropriate scientific-technological infrastructure and a convenient institutional-cultural setting.

The *"users" of available technology* are, in general, those enterprises in the developing countries (LDCs). Of course, there is no level playing field in the LDCs. Some LDCs are relatively more developed and others relatively less developed.

There are several reasons for these developmental differences. The two most important are; 1-) "insufficient" amount of qualified laborers and; 2-) the way technology is "allowed" to be used in the global market. It is unfair to say that there is no technological innovation in the LDCs. From time to time a new technology appears and has a global application. However, these are few and far between. The most "dynamic" industries are; aerospace, genetics, computers and the automotive industry, and technologies in these industries are generally, *under the global control* of the DC enterprises.

The *"creativity"* aspect of "qualified laborer" in a new technology is, often, not necessary in the LDCs. Many *existing* technologies in DCs are, in fact, "new" technologies for the enterprises in the LDCs. Instead of the LDCs "re-creating" an existing technology; looking for ways to maximize the benefits of an existing technology, say, through a technology transfer system seems to be a more rational approach.

However, as mentioned before, at the moment, technology transfer to the LDCs tends to protect and to enhance the interests of the DCs and their companies. In particular, the LDCs suffer from the use of "restrictive clauses" related to the transfer of technology. Furthermore, the so called "transfer-pricing" mechanism contains many defects (imperfections) and works against the interests of the enterprises in the LDCs. There is an urgent need to eliminate these imperfections in the transfer market and the transfer of technology to the LDCs needs to improve (see Gürak, 1990).

Is it possible to Measure the "Qualification Level of the Laborer" Accurately?

Is it possible to measure the mental labor or the qualifications of the laborer, which are of crucial importance in regard to output as well as output growth? If so, what should be the proper method of measurement? Would the outcome reflect the facts accurately?

The issue of measurement bears a great importance for many economists. In fact, according to many economists, if something is not measurable, it cannot be "scientific". The desire to measure is logical; but to expect to achieve definite and repeatable results in the social sciences is illogical. Five factors were put forward as the factors which can affect the level of qualification of the labor force. Mental labor was considered to be the most important factor as it is the foundation of and the eternal source of long-run growth. None of these factors can be measured accurately in order to acquire precise results.

Let us mention again these five factors one at a time: In regard to the first, individuals are endowed with different kinds and degrees of *natural talents*. Some individuals possess talents for sports while some are talented in arts. And there is no technique now nor will there probably be in the future, that could be used to measure such a diverse and multi-level concept such as "talent" in an accurate and universally accepted manner.

The "general" knowledge level varies from country to country. It would not be surprising to find that there is a great gap in the amount of accumulated knowledge between Sweden and Ghana or Pakistan. Sweden has been closely following scientific and technological developments in the past 200 hundred years, while at the same time attempting to establish an appropriate infra-structure. Nowadays, Sweden has many globally competitive firms employing the most advanced technologies, while Ghana and Pakistan seem to be latecomers in these fields. Due to the prevailing conditions in their countries, individuals who grow up in those countries encounter different environments and development levels. Swedish citizens enjoy a wide range of facilities which they can use to learn and benefit from these contemporary advanced technologies, while individuals in the other countries are not even aware of most of these developments. There is no method to measure such divergences and consequently or to make accurate comparisons of their impact on their respective societies.

Formal-informal education and training: To measure the amount of human capital, i.e., the qualifications of the labor force, some economists use the number of school-years attended. Certainly, this can be a method of measurement; but it is definitely not an accurate one. For a period of education lasting 12 years there can be and are considerable divergences in the quality of education and training between different countries. It is also a well-known fact that the quality of formal education or training may vary considerably among the schools in a single country, especially in the developing world. Therefore, the measurement of the human capital, (qualifications of the laborer), based on the "formal school-years attended" criterion can never be accurate.

In regard to the measurement of informal education or training, the solution is no less cumbersome. There are no "school years" to count, nor is there any obvious reason to suppose that informal education achieves better or worse results than the former.

Learning-by-doing & experience: The measurement problem is not any better than in the previous cases. There is no available method to measure the "practice" or "experience" level of individuals accurately. Each individual possesses varying degrees of knowledge and skills, which influences the accumulation of learning-by-doing or experience differently.

Let us recall the initial question in light of all these facts: Is it likely that the level of the qualifications of the laborer which are the source of continual long-run growth can be measured accurately at all? In order to give a positive answer to this question, one has to be either naïve or optimistic. The best outcome would involve measuring "probabilities" or "tendencies".

Other Factors Having an Impact on Growth

The factors impacting on growth are not limited to qualified laborers and technology. Economic policies, social infrastructure, the institutional setting, to name but a few, all effect the rise and fall of a country's growth. Factors such as these only succeed in as much as they create both the environment and the opportunity for growth to occur at a company level. Their contribution to growth is therefore "indirect". We have listed below "other" factors which may impact growth:

1. The institutional infrastructure (the governance, regulations, etc.)
2. The cultural infrastructure (entrepreneurship, tradition, societal expectations, etc.)
3. The local and global framework enabling conditions for fair competition.
4. Economic and political stability.
5. Financial resources.
6. Natural resources.

The Institutional and Cultural Infrastructure

The corporate structure in a country such as the banking system, competition, the foreign trade policy, the influence of non-governmental agencies, the cultural background of the public servants and politicians, its approach to entrepreneurship, labor participation rates, the approach to corruption, problem solving traditional skills, all of these play a very important role in the development of a country.

Mainstream theories which actually ignore these factors are ineffectual in their analyses of growth. Their theories do not contribute much to understanding "accurately" how actual economies function, in neither the DCs nor the LDCs. Unfortunately, these mainstream models, which disregard the "other" factors are still being taught in learning establishments around the world as "credible" scientific theories.

As we see in Russia and the Ukraine, if you have a highly qualified labor force but do not have either the appropriate institutional and/or cultural infrastructure, or the competitive companies necessary, you are not able to meet your targets. The development of an appropriate entrepreneurial and cultural environment takes time. As we are all aware the former Soviet Bloc foundered on these particular issues.

Although we place great importance on the significance, the impact and the contribution to growth from the cultural and the institutional framework, we will not go into them deeply at this time. They will be discussed in more depth in a future publication.

Factors such as the local economics, ensuring *fair competitive conditions*, technological development and the financial structure are a domestic issue and should be dealt with by the decision-makers and rulers of these countries. To claim that LDCs, such as Turkey, are generally not well governed, is not an exaggeration. If Turkey was well governed, literacy would be almost 100 per cent; corruption zero; and there would be no slums. Unfortunately this is not the case.

Competitive Conditions

Factors such as the local economic settings ensuring *fair competitive conditions*, technological development and financial structures are the domestic problems of the countries and should be dealt with by the decision-makers and rulers. Claiming that developing countries like Turkey are generally not governed very well would probably not consider as an exaggeration. For instance, if Turkey were governed well, the number of the illiterate people would be close to zero; Turkey would not be among the countries in the world fraud ranking, and there would not be a slum problem in the urban areas.

Economic and Political Stability

Economic and political stability, in today's world where international interdependency and social sensitivity are paramount, is a hugely important factor affecting growth. Instability decreases the likelihood of positive expectations, development and sustainable growth. It would be ludicrous to expect investment flows in countries with severe socio-economic and political problems as in Palestine or Syria at present (June. 2013).

Financing the Investments

This is another topic closely related to growth. Surveys conducted show that the most important problems for many businesses are insufficient credit facilities. However, overcoming this financing problem is easier than dealing with the other "non-financial" problems. If sufficient financial funds are not available in one particular country, funds can always be transferred from other countries.

Natural Resources

In one sense, natural resources are a gift of nature. However, on their own they are not sufficient to sustain growth. Some oil-rich countries are among the wealthiest countries of the world in terms of per capita income. But, Japan, which has scarce natural resources, is also one of the richest countries in terms of per capita income. That is because; Japan possesses some of the most advanced technologies in the world. Turkey, which has rich reserves of Boron, (a non-metallic element used in hardening steel and producing the rods for nuclear reactors, cannot raise enough income due to the aforementioned imperfections in the global technology market.

Chapter-2
Basic Concepts Related to Growth

Productivity- (Productivity) Growth[5] - Qualified Laborer - Technological Progress

Productivity and productivity growth are two popular concepts frequently referred to, not only by academicians but also by the "man on the street" in daily conversations. Though they are the basic concepts of economics, they often mean different things depending on who is referring to it and on the subject matter in question. The concept of productivity for the economist often has a different meaning than when the same concept is used by a politician, the military, teachers or the mayor of a town. There is a huge and substantial difference between a student's productivity and a carpenter's productivity or a musician's productivity. In order to avoid confusion, it is necessary to define the concepts, within the framework of a particular analysis. The concept of productivity growth is of special significance as it implies economic "growth" which is the main subject of this book.

Productivity... But, by which criteria?

The CEO of an enterprise was very fond of Classical music. Someday, a renowned orchestra comes to his neighborhood. The most outstanding part of the concert is Schubert's well-known "Unfinished Symphony". The CEO receives an invitation to the concert but, due to a previous arrangement, he cannot attend the concert. The CEO gives the invitation to an employee an "*expert on productivity* analysis" and says:

"*Go to the concert and represent me and let me know what you think about it*".

The day after the concert, CEO receives an evaluation report from the productivity expert.

"Dear Sir,

There were four oboe players who were idle during the most part of the concert. It would be more productive to get rid of these.

There were twelve violin players. They all play at the same time and play the same musical notes. This is inefficient; reduce the violin section.

---

5   *Productivity increase* and *growth* are used as synonymous.

> *There was a bias towards 16th notes. What a waste! The audience would not differentiate between an 8th or a 16th note performance. Therefore, these highly paid performance musicians should be replaced with trainees who can play the 8th notes and receive a lower wage.*
>
> *The passages performed by the string section are, in turn, repeated by the wind section. This is an unnecessary repetition. If this is stopped, the concert which lasts two hours would finish in half the time.*
>
> *If Schubert had taken these precautions, the 'Unfinished Symphony' would have been finished.*
>
> *I put this forward for your consideration."*
>
> Productivity Expert

According to one definition, productivity shows the ratio of the output to the input. This ratio is denoted by the term "V". So,

V = Output / Inputs

This ratio could be used in terms of 1-) the "physical quantities"; or 2-) the "monetary value". If physical quantities are the issue, then productivity ratio can be shown as:

$V = Q / (L + X_i)$

Q denotes the quantity of output, L the number of workers and X_i all other inputs of production including capital-goods.

Alternatively, the productivity may be measured in terms of "value", or to be more specific, as the ratio of the total "added-value" (VA) produced to the "total cost" (TC) of the input.

$V = VA / TC$ = Added-value / Total cost

If we use ILO, as our definition for growth, then productivity becomes a concept that refers specifically to the "worker's output". Thus, it ought to be measured as the ratio of the added-value to the labor-wage-cost (LWC), which, in fact, refers to the "partial productivity" (PFP).

$PFP = VA / LWC$

The concept preferred by Solow is one of "total factor productivity" (TFP), which, he suggests is the actual cause behind long-run economic growth. Initially by assumption, there were two factors of production; labor (L) and capital (K). In addition, he proposed a residual; a "heaven sent manna-like" factor in production whose origins are "unknown". It drops in, and is the cause of growth. This newly discovered "magical" factor is in fact a rehash of previous Classical models. Solow named this magical factor "technological progress" which is denoted by "A". So;

$$TFP \text{ á la Solow} = \alpha \, (\Delta K/K) + \beta \, (\Delta L/L) + (\Delta A/A)$$

α is the parameter of capital (K), β the parameter of labor (L) and A, the technological progress which accounts what is known as the *Solow Surplus*. The parts of growth which cannot be accounted for in terms of increases in K and/or L are due to the technological progress A. Let's start with describing what we mean by "productivity" or "productivity growth".

Productivity (V)[6]

The concept of *productivity* simply indicates the relationship between the output and the inputs of production in terms of values or quantities. Thus, *productivity* can be defined as a "static" concept which refers to the relationships between "inputs" and "outputs" of production.

At an enterprise level, or synonymously for an entrepreneur, the concept of productivity normally means "profitability". The underlying explanation of this behavior is that enterprises are established with the sole purpose of making profits, preferably the highest possible profits. If an enterprise is unproductive, e.g., not making profits, it loses the basic motive of its existence. Every enterprise engaged in commercial production has to be productive (profitable) for survival.

Before proceeding with the analysis, it would be useful to point out the differences between "productivity", "economic efficiency", "technical efficiency" as well as "profitability" from the perspective used throughout the analysis.

- *Productivity (V)*: Ratio of outputs to inputs.
- *Economic Efficiency (EE)*: maximizing the difference between total "value" of inputs and total value of outputs.
- *Technical Efficiency (TE)*: maximizing the "quantity" supplied while minimizing the quantities employed.
- *Profit Rate (r)*: profit / total cost of production [$\pi / (p^*q + w^*L)$]

---

6   H. Gürak, 2011, Heterodox Economics.

Measuring Productivity: a Static Analysis

As a static concept, the analysis of "productivity" can be carried out from two different points of view:

1- Quantities; and
2- Values.

The Quantitative Analysis

The "quantitative analysis" of productivity contains some inconveniences and difficulties. For instance, assuming only one type of output, e.g., a homogeneous product, there would be no serious difficulties in measuring the productivity per employee, or "partial" factor productivity (*PFP*) in terms of "one given input". But, when heterogeneous outputs or more than one input is involved, this measurement becomes cumbersome, even unlikely. Assume that 110 tires are produced in a plant and the inputs of production are 10 workers, 20 Kg rubber, 100 KW energy and two machines. Partial factor productivity (*PFP*) with reference to employee can be shown as output per employee.

$$\text{PFP} = \frac{Q}{L} = 110/10 = 11 \text{ Tires per employee} \tag{2.1}$$

Or, alternatively:

$$\text{PFP} = \frac{Q}{X_5} = 110/20 \text{ Kg Rubber} = 5.5 \text{ Tires per 1 Kg Rubber} \tag{2.2}$$

Equation (2.1) shows that one employee produces 11 tires, and eq. 2.2 shows that 5.5 tires are produced with 1 Kg rubber (X_5).

But, what if the inputs are more than one? Would it still be possible to measure total or partial productivity? Equations (2.3) and (2.4) give the answer.

$$V = 110 \text{ pieces tires}/10 \text{ employees} + 10 \text{ Kg rubber} + 2 \text{ machineries} = ??? \tag{2.3}$$

$$PFP = 110 \text{ pieces tires} / 10 \text{ employees} + 10 \text{ Kg rubber} = ??? \tag{2.4}$$

It would not be an exaggeration to claim that it is not possible to measure the total input productivity (*V*) in terms of quantities.

$$V = \text{Quantity supplied / Employees} + \text{All other inputs} = ??? \tag{2.5}$$

Let us use some specific products in eq. 2.5a:

$$V = 20 \text{ Tables / Employees} + \text{Quantities of all other inputs} = ??? \tag{2.5a}$$

In short, the quantitative analysis of productivity with two or more inputs of production is not of much value in analysis, while partial analysis with regard to one input only can give us some useful insights.

Measuring Qualitative Differences

The problem of the measurement of productivity in terms of quantities is not limited to the above mentioned problems. Let us reconsider per employee output in terms of the quantities supplied and analyze the productivity of two competitive enterprises. Assume that an employee at the Volvo plant produces five cars while at Seat the number is seven. Would it be a proper statement of fact that the employee at Volvo is more productive?

The answer is again in the negative. Given the qualitative differences, it would look like comparing oranges with apples. To make a proper and fair comparison of productivity, not only the cars supplied but also the technology employed, the quality of the inputs of production and the merits of labor-force have to be the same; that is *homogeneous*. This happens only in the scientific (?) models of the neoclassical heritage.

In short, a quantitative approach to the measurement of productivity does not seem to be enlightening.

Value Analysis of "Productivity"

Given the specific measurement problems of a quantitative approach, it seems less cumbersome and more explanatory to use an approach based on "value" or to be more specific the value added to the product (VA). The value added to the product is the sum of total wages ($LWC=wL=W$) and total profits (π), including interest payments but excluding taxes.

$$VA = LWC + \pi$$

Assume that total non-wage cost of inputs is 80 TL, total wage cost 20 TL, profit 10 TL, and total revenue after sale 110 TL which provides an added value worth 30 TL. This enterprise is said to be productive both at the enterprise level and at the country level.

$$VA = 20 + 10 = 30 \text{ TL}$$

The proportional productivity relation is:

$$V = VA / TC$$

Where V stands for productivity. Even when the enterprise makes a zero profit ($\pi=0$) it can be regarded as productive from the national economy point of view because it pays wages and the VA is greater than zero.

Assume that, given the total costs including wages, the unit price increases, say due to a shift in demand, and total revenue grows from 110 TL to 120 TL. Now the VA produced is 10 TL higher, although the quantity supplied has not changed. The additional 10 TL is in fact "rent", a type of "unearned" income, e.g., not the result of the productive employment of resources.

Measuring productivity in terms of value is not free of problems, either. However, it seems to be less cumbersome, more reliable and a less complex method than to use the quantitative approach. Using the VA criterion, the productivity analysis can be divided into 5 broad sub-categories. The first one is "total factor productivity" (*TFP*):

1- $TFP = VA/TM$ = Wage + profit / Total prod. cost
$= VA/ (OC+LWC) = (wL+ \pi) / (p^i * q^i + wL)$

Partial factor productivities are indicated below:

2- $PFP = VA/L$ = Wage + profit / Per employee
3- $PFP = VA / t$ = Wage+ profit / Unit time employed
4- $PFP = VA/W$ = Wage + profit / Total wage bill
5- $PFP = VA/OC$ = Wage + profit / Non-wage costs

TM, denotes total production costs; *L*, employees; *t*, unit time employed; *W*, ($L*w$) total wage bill; and *OC*, non-wage input costs.

An Alternative Value Measurement

If the value added to the product (*VA*) is replaced with total income (*Y*) in the definition of productivity, would we be measuring productivity or something else?

V = Total income / Total expenditure = $Y : TC$

Or, alternatively;

$$= (p^f * q^f) / (p^i * q^i + wL) \qquad (2.6)$$

According to some economists, eq. 2.6 measures the productivity, which is, in fact, the proportional relationship between income and expenditure i.e., costs. A closer and careful examination would not overlook that what is actually being measured is, in fact, not productivity but profitability. As discussed above, for an enterprise profitability may imply productivity but for the national economy productivity should be measured in terms of the value added where profitability represents only one component of the value added.

Let us take a closer look at eq. 2.6 with the hypothetical data below. Assume that

$Y = 110$ and $TC = 100$.

$$V = Y/TC = 110:100 = 1.1 \tag{2.7}$$

Eq. 2.7 where r denotes profitability, it has exactly the same result as in eq.2.6.

$$r = Y\text{-}TC / TC = \% \, 10 \tag{2.8}$$

In other words, stating that the enterprise productivity is 1.1 has the identical implication as when enterprise profitability is 10 percent. It is substantially the same for an enterprise, though not for the national economy. Eq. 2.9 is just a different version of eq. 2.7 and eq. 2.8.

$$\pi = Y - TC = 110 - 100 = 10 \text{ TL} \tag{2.9}$$

Optimum Productivity

The optimum position implies that productivity is at its peak, i.e., the highest attainable level with a given technology and labor-force quality. This is the highest productivity level. For instance, given the technology and the labor-force quality, the inputs are assumed to be acquired at the lowest cost and the income obtainable is at its highest possible level while the plant capacity used and the supplied quantities are assumed to be at the maximum possible level. There is no waste of any resources. Under the circumstances, both the economic efficiency (EE) as well as technical efficiency (TE) would be at maximum levels along with the size of profits (π) and the profit rate (r).

The critical question is; is it likely for an enterprise to maintain production at an optimum level? The fact is that, generally, 80 to 90 percent utilization of plant capacity is considered a good performance. When capacity utilization shows a trend towards higher level capacity utilization supported by demand, there would be a good incentive for the enterprise either to expand capacity or to set up a new plant with new investments.

"(Productivity) Growth"

It is imperative for enterprises to enter into a continuous search for increased productivity to reduce unit costs and increase profits. Otherwise, more competitive rivals may gain an advantage and attempt to eliminate their rivals from the market which fits the characteristics of the so called "free market" system. This is known as *"creative destruction"* and was pointed out by Marx and other Classical economists, long before being referred to by Schumpeter. Thus, the search for new technologies is an obligation for all enterprises under competitive conditions.

Technological productivity growth can be analyzed in two sub-categories: the value added (*VA*) and the quantity (*Q*).

1- Quantitative growth: or
2- Value growth.

Quantitative Productivity Growth

A quantitative growth indicates a variation in the relationship between the quantities used in production and the quantities supplied. The remarks made on quantitative "productivity" analysis are also valid for quantitative "productivity growth" analysis. In other words, a measurement of productivity growth with regard to two or more inputs would not produce healthy and reliable results. Measurement with regard to one input only, especially in regard to the employees or the unit time employed would be possible, but the outcome is sometimes bound to be dubious, especially in regard to *"new" products* and *"production methods"*.

Let us assume that Q denotes the quantity supplied L the number of employees and X_i all the other inputs of production in the production function below.

$$Q = f(L, X_i) \quad i = 1, 2, \ldots, n$$

The quantitative productivity growth can be defined as the increase in the ratio of the quantity supplied to the quantity of the inputs of production $[V_t = Q_t:(L_t, X_{i,t})]$ in two different time periods.

$$g = V_{t+1} - V_t$$

In ratio it can be stated as:

$$g = (V_{t+1} - V_t) : V_t$$

Assume that a given quantity (Q) is being produced with less input quantities (X).

$X_{t+1} < X_t$; $Q_{t+1} = Q_t$

Or, assume that output supplied increases with given quantity of inputs; or

$X_{t+1} = X_t$; $Q_{t+1} > Q_t$

Input quantity decreases while output supplied increases; or

$X_{t+1} < X_t$; $Q_{t+1} > Q_t$

Assume that the growth rate of quantities supplied is greater than the growth of inputs;

$X_{t+1} / X_t \quad < \quad Q_{t+1} / Q_t$

In all four cases the measurement of the growth is quantitative. However, the results of such quantitative analysis are unlikely to very reliable, unless the quantities of inputs and outputs are homogeneous. Homogeneity of products would inevitably carry us to the utopian world of neoclassical economics which we hope to avoid throughout this book.

Partial Factor Productivity Growth

Though it might seem, at least initially, less complicated to measure the partial factor productivity (*PFP*) this approach does not seem to produce highly sound or reliable results either. For instance, unless the output supplied is assumed to be a homogeneous product, it is highly unlikely to lead to reliable results. The "homogeneity assumption", on the other hand, implies a journey into the utopian world of economic transactions of the neoclassical heritage.

Let us take a closer look at the partial productivity growth of, say "input item five" among the many other inputs of production, where t stands for time. Assume that the quantity required for the fifth input, say energy, declines by 20 percent from 100 KW to 80 KW due to a technological progress. As we see below the new input quantity of energy is 80 KW now, a 20 KW saving.

$PFP^5_t - PFP^5_{t+1}$

100 KW – 80 KW =   20 KW

If input(s) and output were homogeneous, then there should not be any serious measurement problem of partial productivity growth. However, as the number of inputs subject to productivity growth due to technological changes increases, measurement problems would become inevitable. For instance, how can one make a reliable estimate in a situation where energy demand declines by 20 KW while labor input declines by 10 employees due to a new technology?

To summarize, the arguments stated in regard to the static productivity analysis above are also valid for a dynamic productivity growth analysis subject to the technological progress. In other words, making a reliable quantitative analysis of a specific dynamic productivity growth is complicated and cumbersome, if not impossible. Especially, when *new products* are introduced as a result of technological progress, any attempt to measure productivity growth are bound to be sterile because there are no products of a similar quality with which to compare them.

Added-value Based Productivity Growth

Measuring productivity growth based on monetary value, or to be more specific, on a value added criterion for national economies seems to be a much less complicated process and displays less serious shortcomings compared to a quantitative productivity growth analysis. Therefore, from now on, productivity growth will be considered only in terms of the value added (VA) at a national economy level, unless otherwise stated. Since labor is regarded as the only production factor capable of adding value, it would be rational to measure productivity in relation to the number of workers employed (VA/L) or in regard to the cost of wages (VA/LWC).

Is a "Value" Criterion a Perfect Choice?

A productivity growth analysis in terms of value is, certainly, not free of controversy, but it is more likely to produce reliable results when compared to a quantitative analysis. Here are some examples of the shortcomings.

Case-1

Assume that a technological progress in an intermediary sector reduces the input quantities that are required for the final output, cet. par. To be more specific, assume that now 10 KW of energy is required instead of the 20 KW because of an energy-saving technological progress, cet. par. Further assume that the price of energy (p^e) remains constant. How would this development affect the total value added?

As a result of input-saving technological progress the total value added would increase due to a rise in total profits, cet. par. Yet, there is no increase in the total quantity of the supplied final product.

Case-2

The increase in the amount of the total value added might be due to a wage-rise or increased employment, cet. par. Yet, there is still no increase in the total quantity of the final product that is supplied. This new development would cause deterioration in the economic efficiency (EE) and the technical efficiency (TE), though the total of the value added (VA) would increase.

Case-3

Assume that an enterprise with a monopoly or a few enterprises constituting an oligopoly determine the market sale price of a product and decide to increase the market sale price of the product concerned, while demand remains unchanged. Although the quantity supplied has not changed, the total amount of the value added would increase, cet. par., and appear to be a productivity growth in terms of value.

Case-4

Assume that for some reason the value of the national currency is reduced against other currencies. A comparative analysis of income with another country after the devaluation would give the impression that this particular country has suffered some loss from its previous economic position, although the total quantities supplied in both countries remain unchanged.

In spite of the shortcomings stated above, a value-based productivity growth analysis seems less complicated and more reliable compared to a quantitative analysis.

"New Products" and Growth

Technological progress is the single most important feature of economic and social development. One significant contribution of technological progress is the facility to produce "given" products at less unit cost while another and more significant contribution is the introduction of "new products" often accompanied by "new methods of production". This dynamic process sometimes slows down and sometimes speeds up but never ends.

As the products are "new" with different designs and/or new features, it is difficult to compare them to previous products.

An Inter-Country Productivity Comparison

The "value added" criterion is the most used common criterion when comparing international economic indicators. However, these methods can sometimes provide misleading results. For instance, the 0-16 age group in Turkey is larger than in Sweden. Therefore, an inter-country comparison between Turkey and Sweden based on the per capita value added might be deceiving. A comparative study based on per employee value added seems to produce more healthy results.

The amount of the value added is the most common criterion used in the comparison of international economic indicators. For instance, two frequently used criteria are;

1- the per capita value added (*GDP: Total Population*); and
2- the per employee value added (*GDP: Total Employees*)

to measure and compare productivity. However, these methods can sometimes provide misleading results. For instance, the 0-16 age group in Turkey is larger than in Sweden. Therefore, a cross-country comparison between Turkey and Sweden based on per capita value added might be deceiving. A comparative study based on per employee value added seems to produce more healthy results.

Another interesting issue in any inter-country productivity comparison arises in conjunction with direct foreign investments. Let us consider the inter-country productivity in the automobile sector. As is well known, the more sophisticated and high-technology parts are being produced in the relatively more developed countries, while the developing countries are, in general, assigned to undertake the production of the more labor-intensive parts at a lower wage. In such cases, it would perhaps be more rational to make global profitability comparisons of globally operating enterprises because global production is shaped by their global interests.

Concluding Remarks

A producer always aims to maximize his/her "rate of profit" (r) e.g., ($r=\pi/TC$). This is the only "long-run" driving force for any firm or producer though the short or medium term motive may be different.

However, from the point of view of an economist who analyzes macroeconomic issues and trends, the notion of the "added value" (VA) which combines both profits (π) and labor income, e.g., (LWC=w*L), is a more useful tool in his/her analysis. This gives a more reliable and realistic criterion.

Therefore, throughout this book, we shall use the concept of "added-value" to carry out our growth analysis, unless otherwise stated. In other words, productivity growth implies increases in the added-value.

Chapter-3
Growth Theories: Historical Perspective

Growth has been one of the subjects keenly studied by economists for many centuries. We can see these developments with their titles in the following diagram.

Figure: 3-1 Development of growth theories in retrospect

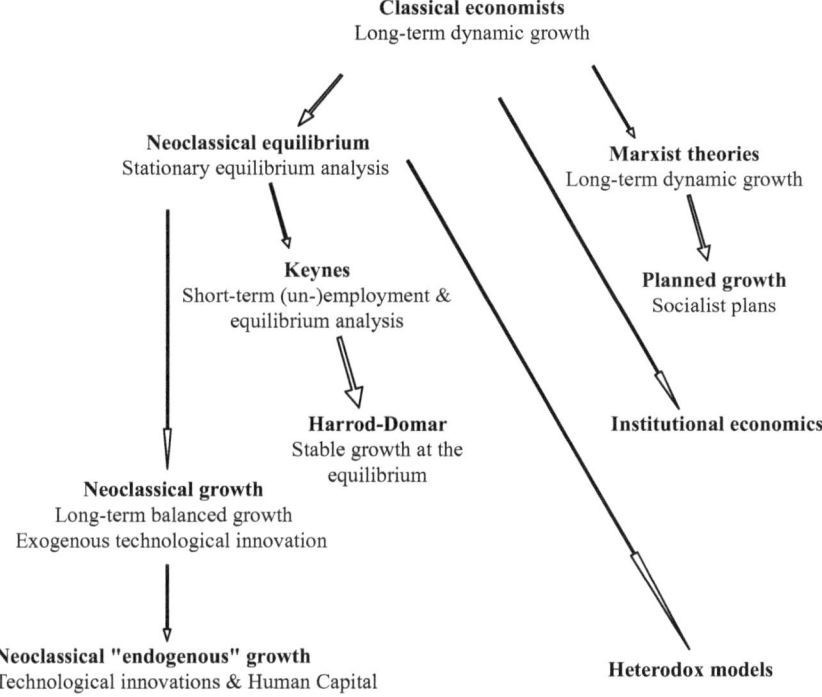

No contemporary economist is expected to object to the contribution of either technological innovation or the laborers' qualifications, i.e., human capital, in the growth process. Almost all of the "new" growth theories already incorporate at least one of these principles of production, and most of the time incorporates them both. Recognizing *technological innovation created by intellectual labor* as the main determinant behind countries' growth adds a new dimension to both value-price theory and the classical definition of capital. In this respect, it is possible to claim that the key input that leads to the welfare growth of a country is 'knowledge', more specifically, the *knowledge required for production*. When we consider that the source of this knowledge is *'creative' intellectual labor*, then *qualified labor* appears as both the most important input of production and a value creating resource. To put it differently, it is the labor force that produces the technology as well as uses it.

Now, let's look at how some economists evaluate technological innovation and qualifications of the labor force in their analyses of growth.

Relationship between growth, new technology and the qualified labor force from a historical perspective.

If we asked a graduate economist or economist-academician in the 1980s questions such as: *"What is the source of welfare of a country? Or; how is growth achieved?* The answer most probably would have been "investments". At that time the common understanding was that, growth could be achieved thanks to investments and the concept that "the investment level is determined by the level of savings ($S=I$)" had been widely accepted. Because money was regarded just as a means of exchange, it was assumed that all the profits earned would be spent on investment ($\pi=I$), therefore, all savings (=profit) would transform into investment ($\pi=S=I$). Even today, there are many economists, politicians and bureaucrats who argue that the growth depends on the level of savings.

In accordance with this $S=I$ approach, when markets get close to saturation point, the demand for products will decrease, the profit rate will fall and, sooner or later, the economy will enter a "stationary" stage. But, the outcome of this pessimistic approach has so far never been realized. On the contrary, despite decreases in the market average rate of profit, new areas of investment have sprung up and continuous leaps have been observed in the profit rates which coincided with increased technological innovations. The cause of these *macro productivity increases* or *technological productivity increases* is the invention and use of *new* technologies, which are the product of *intellectual labor*. Every "new" technology symbolizes a potentially higher rate of profit expectation, new areas of investment and increasing demand. There may still be some economists who do not admit that technological innovation, a product of intellectual labor, is the reason behind the sustainability of long-term growth and the reason why profit rates do not enter into decreasing period towards zero. The widely acclaimed view is, however, presented differently in different models, that;

1- The qualifications of laborers (or, synonymously, the quality of labor),
2- Technological innovation

contribute, in one way or another, to the growth process.

If this is the case: Why weren't those economists who made significant contributions aware of these two important basic factors?

In the past economists, particularly the Classical economists did not generally deny the role of the quality of labor and/or new technology in their analyses but, unfortunately, they neglected or overlooked these factors and therefore they did not give them their deserved and required role in either their value/price or growth theories. One of the main reasons behind this is that they prioritized other factors.

For example Smith and Ricardo, proving the advantages and legitimacy of the emerging infant capitalist order, over the Mercantilist and Feudalist ideologies was their most important issue, whilst for Marx, proving how labor was being exploited was his main objective. In the 1900's the basic aim of the widely acclaimed Marginalist doctrine, was to create abstract 'equilibrium' models consisting of homo economicus-type robots, which were regulated by "scientific (?)" laws but without any input from the historical, political, traditional, or human perspective. While trying to achieve this goal, they did not realize the importance of technological innovation until the 1950s and of the quality of labor (human capital) until the 1960s. They probably neglected these factors in order "not to upset the equilibrium" state of the utopian economic models they had created.

In the substantial part of the recent century, growth models treated technology as a 'given' factor in production and the quality of labor as immaterial other than in one or two exceptional cases. On the other hand, around 2,400 years ago the well-known philosopher Hippocrates pointed out the significance of the human brain and its impact on daily life. Humans should know that joy, cheer, laughter and joke as well as sadness, grief, shriek and wail come from nothing but brain. With our brain and attitudes, we gain intelligence and wisdom; see and hear; have the reason to know to what is and distinguish right and wrong, good and bad, tasteful and tasteless.

Unfortunately, the human brain, an extraordinary and to some a visually unpleasant organ whose impact on changing our life and surroundings, has been neglected by economic theories for many years. It is evident that the relationship between the brain, intellectual labor and production is not at its "should-be" level in the fundamental subjects of economics such as *"value/price"* or *"growth"* theories. University economic academicians have been allowing the neoclassical interpretation of price and growth to dominate their teaching. This practice of putting forward impractical and useless theories serves no-one, save for the purposes of self-aggrandizement. Unfortunately, today, many economists still treat the neoclassical doctrine and its position in economics as if *'they have reached the peak of their endeavors in their quest to possess permanent and unchangeable universal laws*' and perceive this utopian economic world of theirs as if it were real.

It is accepted today that the science of economics began with Adam Smith around 230 years ago. Now, let's look at what position the quality of labor and new technology was in the economic thinking during Smith's time and what developments have occurred since.

A. Smith

As the founder of the science of economics, Adam Smith's ideas should be evaluated in the light of the facts and knowledge available at that time. Otherwise, we will treat him unfairly as well as misinterpreting the message he wanted to convey.

We should remember that during the period in which A. Smith penned his works a transformation from Feudalism to Capitalist was taking place. The ideas of the physiocrats were widely acclaimed, and international economic relations were dominated by the Mercantilist ideology. Smith was, in a way, acting as an ideological spokesman for the emerging capitalism and demonstrating reasons for changing the society. That is because, the guild system allowed a limited level of production. The division of labor in the production process, as suggested in the 'Wealth of Nations', would increase productivity. The economic system had to change and be restructured. In order to achieve this, concepts such as production, labor and capital had to be redefined according to the emerging conditions.

The time when Smith endeavored to build an ideological infrastructure suitable for the emerging processes was technologically dynamic and many "new" inventions were being introduced. Thanks to these technological innovations, in addition to the dexterity in every particular workman and to the saving of time, productivity was increasing on the one hand, profitability on the other.

According to A. Smith, the production of commodities (the physical goods) was subject to *the law of increasing returns*. As we recall from the previous sections increasing returns are a consequence of the technological innovation. However, A. Smith did not attempt to establish a direct relationship between the wealth of nations and technological innovation. He seems to regard "technological innovation" as a consequence or a derivative of the division of labor which inevitably led him to formulate a 'limping' wealth growth model. According to A. Smith technological innovation followed a 'division of labor' which perpetuated a growth in wealth. Therefore, while the influence of technological innovation remained in the shadow, capital accumulation and the division of labor took center stage. All "new" machines and equipment leading to the increase in productivity seemed to be inventions that followed the "division of labor" (Smith, 1985; 22)[7].

When it comes to the issue of qualified labor or the quality of labor (e.g., human capital) we know from his various works how important education in the new economic order was. As an economy grows quickly, the demand for the qualified labor grows. To meet this demand, it was necessary to enhance the quality of the laborer. According to Smith, educating an individual should be regarded as a kind of investment leading to future returns. He particularly advocated for a free education service for children of poor families as it would be significant in terms of the future of the system.

In his first Book, Chapter 5, by defining that the labor used is *"the real criterion for the exchange value of all commodities"* it shows that Smith was aware of the fact that labor with differing levels of qualification will have different levels of productivity.

> *"... though labour is the real measure of the exchangeable value of all commodities, it is not that by which their value is commonly estimated. It is often difficult to ascertain the proportion between two different quantities of labour. ... There may be more labour in an hour's hard work than in two hours easy business; or in an hour's application to a trade which it cost ten years labour to learn, than in a month's industry at an ordinary and obvious employment."*
>
> (A. Smith, 1976; 48)

Although Smith stressed the importance of education that would supply the necessary qualifications for the labor-force that would increase the national wealth, and also noted that differences in the level of qualification would lead to differences in productivity, he did not emphasize sufficiently the creative role of intellectual labor.

In fact, the primary building block of the division of labor, suggested by Smith as the root of the nation's wealth, was *qualified labor*. Smith's example on how the division of labor increased the productivity in pin production is very well-known by every economist. As we explained in the previous sections, the thought of the "reorganization" of the division of labor was a result of the intellectual assessment made by an entrepreneur or by a creative minded employee. Therefore, the factor which leads to increased productivity (growth) based on the division of labor is the intellectual assessment made by a qualified laborer

---

[7] "... the invention of all those machines by which labour is so much facilitated and abridged, seems to have been originally owing to the division of labor." (Smith, 1985; 20)

who performs the reorganization of the production. The entrepreneur, as a result, prefers the division of labor since he expects to earn higher levels of profit. A growth in productivity which is sustained due to intellectual assessment is a tool and the intellectual labor; the original source of productivity growth. In other words, productivity growth, in fact, stems from "the productive knowledge", which is a product of the human mind. However, Smith, unfortunately, did not establish a concrete theory based on the qualified labor-growth relationship. In fact, in a surprising and interesting approach, he argued that there was not any human basis at the root of the division of labor and accordingly of the resulting positive changes. Thus, the division of labor, from which so many advantages are derived, was not considered as the original effect of any human wisdom (A. Smith,1976; 25). As Adelman (1972; 27) pointed out, technological innovation in Smith's model *"happens by itself"*. Making use of these innovations is *"possible with capital accumulation"*.

Therefore, in Smith's works, neither technological innovation nor a qualified labor force could be the necessary focus of attention as they deserve when defining the context in which growth increases. The reason for this neglect could be blamed on the ideological struggles and pool of knowledge and experience existing in Smith's time. Yet, when Smith is evaluated in the context of his own time period, his very important contributions to the science of economics is undisputable.

If we are to summarize Smith's *"endogenous"* growth theory:

- *There is technological innovation.*
- *The basic source of growth is the 'division of labor'.*
- "Division of labor is *not a conscious product of thought* which envisages and aims at general wealth (Smith, 1985; 25).
- Since productivity will increase as a result of the division of labor, the *law of increasing returns* per capita employee prevails.
- In the long-term, the profit rate will fall and growth will halt.
- The education/training of the labor force and thus its level of qualification is very important. He defends free education for the children of poor families.
- He does not attempt to establish a relationship between the institutional-cultural framework and growth.
- *He does not establish any relationship between creative intellectual labor and growth; however, attempts can be made to draw a relationship between 'learning-by-doing' and the growth process.*
- *The relation between creative intellectual labor, technological innovation and growth is almost non-existent.*

Ricardo

One of the classical economists who contributed significantly to growth theory is Ricardo. Like Adam Smith, Ricardo was interested in subjects such as value and price theory, wage rate, profit, exchange relations, international trade and growth and had also been in pursuit of a "factor" of production in which value is an invariable. In addition, he placed great importance on functional income distribution and although he believed in technological innovation and the law of increasing returns in industry, he argued that income distribution would change in favor of the landowners.

In Ricardo's time, in England, industrial production and employment was skyrocketing thanks to new investments. Technological innovations were continuously creating new opportunities for earning profits while demand for agricultural products was increasing. However fertile agricultural land was decreasing. Under these circumstances, according to Ricardo, it would be impossible to sustain a continuous increase in industrial production. Sometime or other, increasing costs in the agricultural sector would increase the real wage level, decrease profit rates and eventually economic growth would come to an end.

We can explain Ricardo's model with the help of Figure: 3-2. Two concepts are important in this model which explains income distribution, as well as growth:

1. rent (for the agricultural sector) and
2. the surplus value or the added value (wage + profit).

Figure: 3-2 Ricardo's model

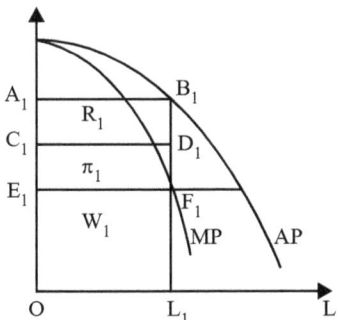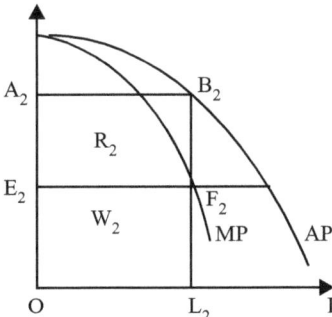

(w) denotes long-term wage, (L) employment. When employment is OL_1 then total product will be the area of $OA_1B_1L_1$. The share of labor is the area of $OE_1F_1L_1$, share capital the area of $C_1D_1E_1F_1$ and income the area of $A_1B_1C_1D_1$. As a result of a productivity increase (growth), the profit will be zero when demand rises from

L_1 to L_2 while growth will end and the income will only be shared between the laborer ($OL_2E_2F_2$) and the renter ($A_2B_2E_2F_2$). Since, in the long-run the wage-rate will be at the subsistence level, income distribution will deteriorate to the detriment of the capitalists but in favor of the landowners.

As seen in the Table: 3-1, when rent is zero and the most productive agricultural lands are employed (Y_t). At the next stage less productive agricultural lands are in use, the rent increases over time and it reaches its highest rate at Y_{t+3} at which point the income of the capitalists is completely lost. Real wages were high at the start, but over time, they would decrease as a result of population growth and an increase in the prices of agricultural products.

Table: 3-1 Long-term income distribution

	R	W	π	Total
Y_t	0	50	50	100
Y_{t+1}	20	40	40	100
Y_{t+2}	40	35	25	100
Y_{t+3}	70	30	0	100

Although Ricardo, like A. Smith, defends the principle that growth is rooted in investments, he did not attempt to establish a theory showing the relationship between growth and "technological innovation". Yet, Ricardo believed in a continuous development of new technologies by virtue of competition in the industry and the prevalence of the law of increasing returns. The reason why he did not make any effort to further develop this relationship may be attributed his other priorities at that time. For example, while he was criticizing the errors that he found in Smith's model on the one hand, Ricardo was trying to postulate a better value-price theory using the same framework. In addition, he was looking for a value criterion that was invariable, in order to make objective measurements, using the rent hypothesis, to demonstrate that landowners have a greater advantage in terms of income distribution in the long-run. By taking a negative stance that increasing returns due to new technology in the industrial sector is not applicable to the agricultural sector and in the long-run "the law of decreasing returns" would prevail and sooner or later economic growth would come to an end. However he underestimates the effects of technological innovation.

In fact, Ricardo's primary aim was not to analyze the growth process. That is because, according to him, the aim of political economy is not to study the structure and reasons of wealth, rather, to analyze "the laws that determine income distribution" among the classes. Furthermore, when Ricardo refers to the quality of labor, he uses the concept of "different labor types". However, Ricardo discussed the concept of "different types of labor" within the framework of determining the "value" of commodities. He did not analyze the "different types of labor" concept in relation to technological innovation and its effects on the formulation of price. Therefore, like technological innovation, the "quality of labor" concept could not find its deserved place in his model in terms of its contribution in growth. As noted by Kaldor, Ricardo's economic analysis based on the Law of Decreasing Returns inevitably leads to a *"stationary equilibrium"* in the long-run (Kaldor, 1986; 190).

The general characteristics of Ricardo's *"endogenous"* growth model are:

- The short term wage rate is $w=f(N_d)$; and the long-run wage rate is at subsistence level.
- Population growth is "endogenous". $N_s = f(w/p)$
- Savings are "endogenous"; $s = f(\pi)$ and $S=I$ (π=profit)
- There are two production factors: K=capital and L=laborer.
- Full employment.
- Perfect competition conditions prevail.
- Say's Law prevails.
- No attempt to establish a relationship between the growth process and the institutional and the cultural framework.
- *Technological innovations arise from competition.*
- *"The law of increasing returns" prevails in industry.*
- But "the law of decreasing returns" prevails in the agricultural sector and the overall economy.
- *There is NO analysis establishing the relationship between the quality of labor and growth.*
- *Technological innovation does not secure the long-term growth.*

Marx

According to Hicks, Marx is the last representative of the old school of growth theory (1966; 263). Marx stands out amongst the classical economists because of the importance he placed on technological innovation. His growth model consists of investments (I), population growth (n) as well as technological innovation (A).

$g = \Delta Y/Y = f(I, n, A)$

However, Marx's interest in technology was not about how technological innovation impacts on growth, rather, how it augments the surplus value gained from the laborer, which is the source of wealth. Basically he was concerned about how laborers were exploited. His theory of growth was in general established on the surplus value and the amount of investment. The aim of the technological innovation, a requirement in order to increase competition, was to increase the productivity of the labor force, therefore, the surplus value. Increasing the productivity of the laborer meant less employment opportunities for the labor force per unit of production time and therefore decreased the value or price of the commodity produced. Because, the value of the commodity was determined by the abstract concept of the labor-time employed (Marx, 1977; 303).

New technologies caused productivity growth, but since Marx did not establish a relationship between the "new technology" and the "quality of labor", the result was the contributions of these two important factors in both growth and the addition of value to a commodity was ignored.

There are three ratios determining the growth rate in Marx's model:

1- Surplus value ratio; s/v
2- Profit ratio; $s/(c+v)$
3- The organic composition of the capital employed; c/v

where, (s) denotes the wage deficit that the worker suffers (the surplus value), (v) the wage paid to the worker, (c) the fixed capital. Profit rate now determines the growth rate.

$$I = f(r)$$

But the profit rate, by itself, depends on the surplus value rate i.e. the rate at which the workers are exploited (exploitation rate) (s/v) and the organic combination of the capital (c/v).

$$r = s/(c+v)$$

The surplus value ratio (s/v) is assumed to be constant. The organic composition of the capital (c/v) will increase due to technological innovation over time and this will lead to a decrease in the rate of profit. In other words, fixed capital investments (c) will increase due to competition over time and since the exploitation rate (s/v) is assumed to be constant, the profit rate will decrease. When the profit rate is zero, no new investments will be made; thus the "effective demand" will decrease and the economy inevitably result in a depression.

Example:
At the beginning, let's assume s=100, v=100 and c=200.

$$r = s/(c+v) = 100/200 + 100 = 100/300 = \sim 33\,\%$$

Let's assume that, as a result of new investment due to competition, the fixed capital amount increases by 100 units; $c' = 300$, cet. par. In the new case, profit rate will be lower.

$r' = s/(c'+v) = 100/300 + 100 = 100/400 = 25\%$

A continuous rate of decline in profit will cause investments to decrease. This leads to a decrease in "effective demand" which means that an economic depression will be inevitable. If depressions did not exist, the only obstacle for growth would be the unavailability of laborer.

In Ricardo's analysis, an increase in rent causes the profit rate to decrease, while in Marx's model the profit rate decreases as a result of an increase in the organic composition of the capital (c/v). In other words, according to Marx, "new" technologies which require a higher fixed capital expenditure indirectly cause a decrease in the profit rate.

In reality it is just the opposite. New technology leads to an increase in the surplus value per worker. As a result, although fixed capital investments increase over time, the profit rate does not decline towards zero and the long-term economic growth rate does not drop towards a stationary situation. The main reason Marx could not foresee this development was partly due to his belief that new technology augmented the exploitation rate thereby proving that exploitation of the workers actually existed. If he had analyzed technological innovation in the context of growth, his conclusions would probably be very different.

In the post-Marxist period, Marxist economists intensified their efforts to defend Marxism ideologically. In time, even they began developing pure scientific (?) models similar to those of neoclassical economists. In fact, the models they developed were as "pure" "academic" and sophisticated as the models developed by the Neoclassical School, however they were unrealistic.

However, in his analysis of capitalism, labeling its characteristics as 'creative destruction', Marx has left his supporters an important clue about the economic impact of technological progress. If Marxist economists had analyzed the causes and results of technological innovation with due process and how it affected and changed output and society, they could have better explained the economic growth process and why profit rates do not tend to fall towards zero. Interestingly, it wasn't the Marxist economist that adopted this approach but an Anti-Marxist economist named J. Schumpeter, whose analysis gained him quite a reputation.

Marx was aware of the importance of the "quality of labor" but he did not put so much emphasis on demonstrating its contribution in the value/price formation and in the theory of growth. We may even say that he ignored it. According to Marx who used the use- and exchange-value concepts developed by Ricardo, value of a commodity was determined by the "labor-power". "Labor-power" was the aggregate of the *intellectual and physical skills* of an individual[8] (Marx; Vol. II; 270). Concrete labor with different qualifications produces the use-value while the exchange-value is determined by abstract labor. Being aware of the differences in labor-qualifications, Marx had preferred to use the concept of abstract labor which can be qualitatively calculated as the "labor-time employed in production". Therefore, it inevitably became difficult for him to establish a relationship between the qualified laborer, technological innovation and the growth process.

Marshall

One of the most reputable economists of the science of economics, Marshall did not develop a theory of growth. However, we will mention about him, because of his important contributions in the economic analysis and having a reputable place in the economics. If Marshall's warnings, as a person who placed great importance on the education and the knowledge in general, had been taken into account, the science of economics would have probably undergone a very different course of development.

According to Marshall, "general education", which helps individuals to gain qualifications, is very important because it increases the productivity. A good general education enhances the intellectual capabilities of a worker; it gets them adopt rationalist curiosity habit; enables them to be more clever, more prepared, more reliable; and enables them to have higher quality of live during working hours and in leisure times. All these would improve their level of wealth (1961:211).

According to Marshall, technical education which also includes craftsmanship education is also important because it enables the development of control on the eyes and fingers. But, the general education helps to develop the "creative" capability of the individuals. (1961:209). When allocating the resources to education, we should not only consider today but also tomorrow, says Marshall. That is because he believes in many direct or indirect returns of the education in the long-term.

Compared with the Classical economists, having carried the flag one step forward in the issue of knowledge, Marshall set forth the idea of *"knowledge is the most powerful engine of the production "*(1961; 115). In here, what is meant by

---

8   "... the aggregate of those mental and physical capabilities existing in the physical form, the living personality" (Marx, Vol. II; 270).

"knowledge" is, of course, the knowledge required for production, namely, technology and it is materialized in the machines. Machines enhances the power of human being on the nature as well as his/her control on his/her own life and the surroundings and facilitates his/her life by enabling him to control and change his/her surroundings. Therefore, machines mitigate the workload of human beings and saves human from heavy works requiring muscle power (1961:262).

According to Marshall, as the civilization makes progress, humankind produces "new demands" and "new methods" of production. Everything develops very fast and it is unknown where this process will come to an end. Therefore, according to Marshall, there is not even a sign that we are at a point close to the "stationary equilibrium" as described by the Neoclassical doctrine (Marshall, 1961,185).

Not only new technologies released human beings from heavy works but while reducing per unit production costs, thus improving the competitive advantages of the companies, also offered cheaper products of consumption. In addition, new technologies contribute in the economy through internal as well as external economies.

In his work, Marshall clearly states that production has "only" two factors; *nature and human*. The means of capital and organizational structure argued by some economists as "productive" are the produce of the laborer (1961; 116). In other words, the productive labor is rooted in the value of exchanged products. In this approach of Marshall, we observe that his idea of value has some similarity to the Marxist economic thought, but root of which dates back to much older times.

According to Marshall, as a result of development, qualified labor became much more important in the production and the technological innovations provide the means which will change and make our lives easier, besides increasing control on our surroundings. However, having approached so closed to *qualified laborer-technological innovation-growth* triangle, Marshall was also not in the pursuit of a theory that will demonstrate the relations among these three key factors. Albeit, as pointed out by Hicks, two chapters of Marshall's "Principles" (1961) are about "economic progress"; but these are sketchy things penned in urgency (Hicks, 1966,258).

To summarize the ideas of Marshall:

- *Technological innovations exist and they are continuous.*
- Due to the continuous nature of technological innovations, it is unlikely to have a "stationary equilibrium". In other words, there seems to be *no limit to the growth process.*

- He does not attempt to establish relations between the institutional and cultural structure and the growth process.
- In *growth process*, education, therefore the *qualified laborer is very important*.
- There is no theory based on the relations among the qualifications of laborer, technological innovations and the growth process.

Keynes

The role of technological innovation in the growth theory was still ignored in Keynes' analysis. It would not, in fact, be quite fair to claim that Keynes had no theory at all in regard to growth. His ideas to reduce involuntary unemployment by government intervention can be regarded as a 'short-term' theory of growth. However he provided no long-term growth analysis.

The economic models developed by Keynes' followers have a significant position in macroeconomics textbooks. But, in fact, the measurements which are expected to eliminate involuntary unemployment do not target long-term growth specifically, but rather aim to restore a state of "full employment equilibrium". The so called "Keynesian" measurements to restore full-employment are in accord with the definition of 'short-term' growth, as has been outlined in the Chapter-2 of this book.

When evaluating Keynes' "General Theory" and the ideas of his followers, it is useful to approach the issue from two different perspectives.

1- In terms of economic theory.
2- In terms of economic policy (implementation).

One of the basic goals of Keynes was to criticize some of the basic theoretical approaches of the neoclassical doctrine and to demonstrate that equilibrium is also possible with some level of unemployment. According to Keynes' predictions, when a state of involuntary unemployment exists and the autonomous investments are insufficient to increase the output and employment, then governmental intervention would increase the effective demand, thus further increasing employment as well as the output. Eventually the Neoclassical full-employment "equilibrium" would be restored.

The purpose of Keynes' model was not only to demonstrate the possibility of underemployment equilibrium in an economy but also to demonstrate the role of money, expectations as well as the inelasticity of the wage-level. In fact, Keynes did not entirely reject the neoclassical full-employment equilibrium model; he merely pointed out the necessity of looking at the model from different dimensions. Keynes was especially concerned with the short-term cyclical fluctuations and did not present any ideas about the long-term growth process. Moreover, since he did not deal with the quality of labor or technological innovation, he did not attempt to establish a relationship between them and the growth process.

It would be unfair to attribute the practices of fiscal policies to invigorate the economy supported by Keynesians solely to Keynes' ideas. Because, as we very well know, these fiscal policies had been known and implemented in various countries at least 70-80 years before the publication of Keynes's book. The fiscal policies to rescue an economy from depression and reinvigorate it had already been in place and were successfully implemented in the USA during 1930s and in Germany during Hitler's time years before Keynes penned his book. Therefore, claiming that the ideas on fiscal policies implemented to restore the economy and full-employment were developed by Keynes leads us to treat the USA and German economists unfairly.

Keynes's Model

The essence of the General Theory is based on the employment issue which depended on:

1. The aggregate supply function.
2. The propensity to consume.
3. The volume of investment.

The total income of a country is determined by the amount of employment. At full employment level, the total income will be at the highest optimum level. If the unemployment prevails, the output, thus employment has to be stimulated by increasing the effective demand (ED^9).

Let's assume that at the beginning unemployment equilibrium exists at point *D1* (see Figure: 3-3). Until reaching the point D^* which is the full-employment equilibrium point where all the factors of production are employed, the supply will be elastic and accords with the effective demand ($ED=Y=C+I$). Therefore, every increase in effective demand will also increase/ supply. If the marginal efficiency of capital (*MEC*), i.e., the rate of return expected, is lower than the interest rate (i), ($MEC<i$), then autonomous investment will not be made. In this case, only the governmental expenditure (*G*) will produce the expected result i.e. to restore the full-employment equilibrium (Y^*).

---

9 Effectice demand: the sum of expected consumption and the expected investment.

$Y* = f(C + I + G)$

As it is seen, Keynes's model does not embrace either technological innovation or the quality of the labor. Because the purpose is not to analyze the long-term growth process, but to demonstrate how full-employment equilibrium will be restored from the un-employment equilibrium in the short-term with certain 'given' inputs. In other words, Keynes's approach analyzes the cyclical fluctuations with a *'given' technology and a 'given' level in the quality of the labor*. Since there are no new products and no new production methods, the stationary period starts after reaching the point of equilibrium.

Figure: 3-3 Short-term growth and equilibrium

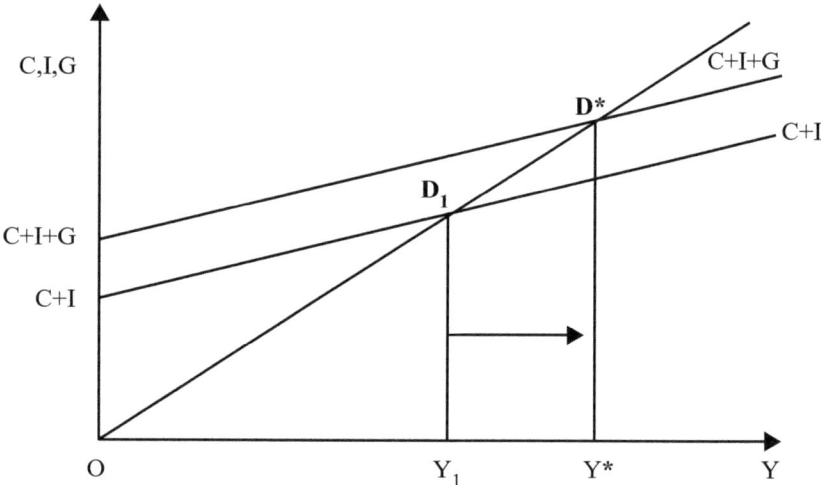

We can analyze the equilibrium state without growth, excluding population growth, from a different point of view with the help of Figure: 3-4. If the economy is at equilibrium at D_1, it means that unemployment equilibrium may exist, as described by Keynes. At the point D_2 over-employment occurs. In both cases, the economy has to move in the direction/ shown by the arrows in order to reach equilibrium $D*$, namely the equilibrium production level $Y*$.

Figure: 3-4 Equilibrium and growth

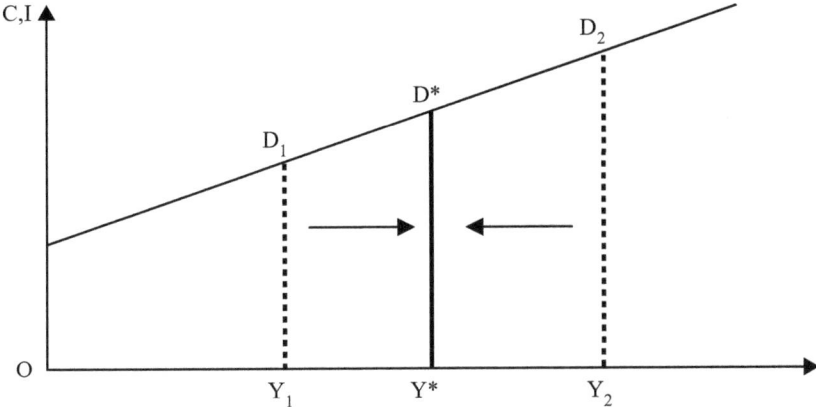

The Characteristics of Keynes' "Static" Model

- "Equilibrium" is also possible with un-employment.
- Wage level is not determined by the supply-demand law and the wage level is not elastic; in fact, the wage level is resistant against decrease.
- Say's Law does not prevail. Supply is infinitely elastic but the quantity supplied is determined by the effective demand.
- Interest rate is determined by the supply-demand conditions of money, not by the amount of money. [$I = f(M^s, M^d)$]
- Expectations are significant.
- If $I = f(MEC, i)$; $MEC > i$; investment continues.
- When $MEC=i$, then equilibrium is reached.
- If full-employment is not reached and the autonomous investments are insufficient, then governmental expenditure (G) is required.
- No analysis on the relationship between technological innovation and growth process.
- No analysis on the relationship between the quality of labor and the growth process[10].

---

10 According to Harrod, Keynes did not revolutionize the theory of economics. The contribution of Keynes is a re-adaptation of mainstream theory and enabling a shift in the focus of subjects.

Harrod-Domar

It is hard to imagine a book about growth theory without the ideas of Harrod and Domar. Their basic approach, which we can regard as a follow-up of Keynes' model, was to point out the basic conditions required for 'steady growth'. Because of the risk that ex-post investments do not equal the ex-ante investments, demand was expected to increase along with an increased production capacity, but growth would be unstable. Therefore, the conditions for steady growth have to be determined, so they attempted to calculate what the optimum growth rate should be.

Assumptions of the H-D model:

- Homogenous commodity, i.e., the SAME commodity is produced and consumed.
- Two factors of production; labor (L) and capital (K).
- Constant returns to scale (*CRTS*).
- Growth is a function of the savings $g=f(sY)$
- $I = S = sY$ (exogenous)
- K/L (constant)
- $\dot{L}/L = n$ (exogenous)
- No technological innovation.
- No quality of labor.

At equilibrium, ex-post investments have to equal to ex-ante savings:

$$I_{ex\ post} = S_{ex\ ante} = sY$$

For a steady growth:

$$g = s/v = n$$

($s = I/Y$; $v = Y/K$) but since (seven) are three independent variables, equilibrium is not guaranteed.

At the beginning, let's assume that $g=n$, but for some reason, $I_{ex\ post} > S_{ex\ ante}$. With the effect of the multiplier, total income (Y), will grow faster than the capital stock ($\Delta Y/Y > \Delta K/K$). As a result, the capital to total income ratio (K/Y) will decrease, new investments will become more profitable and production capacity will increase. Consequently, since the equilibrium has broken down due to $I=S$ inequality, instability will occur.

As has been shown, the Harrod-Domar model did not take into account either the role of "new" technology, or the quality of labor, nor does it explain their contribution to the growth process.

Having argued that full-employment and steady growth is not likely to occur, Domar denied his own growth model around 11 years later the World Bank and some international organizations still continue to use the "capital to output ratio- (ICOR)[11] in their calculation of growth. For example a report of the EBRD[12] for the year 1995 stated that a 20 percent saving (ICOR=4) is required for a 5 percent growth rate, in respect to a former communist country. The World Bank again in a report for 1995 suggested that an 8 percent rise in investments i.e. savings, will increase the growth rate by 2 percent (again ICOR=4) (Easterly;1998).

According to the model used, the factors which determine growth are the magnitude of investment and the incremental capital output ratio (ICOR). However, in such a model where technology is regarded as a 'given', growth may only continue until the markets reach saturation and then if sufficient levels of purchasing power exist, it may continue to the extent of the population growth.

A summary of the Harrod-Domar model

- They tried to develop a steady growth model at the equilibrium level "a la" Keynes. But after years Domar admitted that their efforts were not satisfactory.
- *No analysis on the relationship between technological innovation and growth.*
- *No analysis on the relationship between the quality of labor and growth.*

Schumpeter

Schumpeter's ideas on technological change seem to be substantially inspired by Marx. He focuses on the fact that capitalism which inherently and *constantly* undergoes revolutionary change, destroys the old production relationships for the sake of constantly creating new ones (Schumpeter,1970; 83). However, technological innovation also disrupts stability. As supply increases, the products to be consumed also increase, but input becomes/ scarce. As in Ricardo's model using the diminishing fertility in agricultural lands, as industrial input gets scarce, prices, therefore, their production cost will increase. Increasing cost means decreasing profit therefore a decreasing level of investment. Another problem which causes the profit rate to fall is the increase in the demand for loans in order to increase output which results in higher interest rates.

According to Schumpeter, the impact of technological innovation on growth is important. Although Schumpeter was not looking to developing a theory

---

11 ICOR: Incremental Capital Output Ratio.
12 EBRD: European Bank for Reconstruction and Development.

of growth, he highlighted a very significant factor i.e. *new technologies are not manna from heaven* as in the neoclassical theories of growth imply; *they emerge as an inevitable and inherent result of the system.*

Schumpeter's ideas on the impact of technological innovations on the long-run growth and the concept "creative destruction" were not original, though some economists suggest they were. In fact, we have previously encountered similar ideas in the theories of Marx and in Marshall's works, for example, long before Schumpeter's book. Obviously Schumpeter was inspired by the works of Marx and others on the role of technological innovation. Anyway, it was a brave and realistic attempt by Schumpeter to emphasize the significance of the role played by technological innovation on long-run growth and its impact on competing firms when the neoclassical economic ideology was in its heyday.

To summarize;

- *Technological innovations are endogenous.*
- *Continuous technological innovation drives growth.*
- Creative destruction' is the characteristic feature of capitalism (Schumpeter, 1970; 83) and the basis of this characteristic is the technological innovation.
- In a period dominated by *equilibrium* models and doctrines, the importance that he placed on technological innovation and the interest that he attached to the subject is important. But his ideas on this subject are not original.

Overview

As we clearly have seen in this section, technological progress was not an unknown or unexplainable event for the renowned pioneers of the Classical movement like A. Smith, Ricardo, and Marx in the 18 and 19 Centuries. In fact, they were well aware of the crucial and significant role played by technology. In other words, things were moving as they should and the theories/models were not only:

1- Logical and;
2- Consistent, but also;
3- Based on realism.

However, since the 1870s, economic science began undergoing ideological changes. Some ideologues supported by the concerned social classes started introducing "new and alternative" ideas to Marx's radical analysis. Economic science was gradually being transformed into a "pure science" with "universal laws" as in the natural sciences such as astronomy and physics. As time passed, economics lost all its affiliation with historical developments and/or normal human behavior or enterprise activities.

In the meantime "endogenous" technological progress was ignored totally and the static steady-state analysis determined the shape of Global intellectual thinking. Certainly now and then, there were some exceptional views like those of J. Schumpeter who seems to be the most renowned economist regarding the role of technological progress in growth process.

However, one day a miracle occurred and R. Solow shouted: *EUREKA !*
Technological progress was "re-discovered".

Chapter-4
New Approaches to Growth Theory

We started by analyzing growth by emphasizing the importance of a "qualified labor force" and "technological progress" in Chapter 2. These two key factors had unfortunately been neglected by economists for a long time with the advent of the Marginalist school of thought which emerged in the 1870s. The primary aim of the proponents of Marginalist School was to confront and eliminate the Marxists threat. Instead of attempting to develop rational alternative theories, they introduced fictitious "equilibrium" growth models with fictitious assumptions. After equilibrium further growth was subject to "exogenous" population growth rate. Technology was considered to be a "given" factor. Questions such as: Where does technology come from? How does it impact the production process? were paid lip-service only. The assumption that "laborer and capital goods are homogenous" only made the situation worse. As a result, instead of growth, cyclical fluctuations became the focus of economists.

After the 1950s, we observed a progressively increased interest in the issue of growth as a fact as well as in the sources of growth. Thanks to Solow, the importance of technological innovation was "re-discovered". But because it was assumed to be an "exogenous" factor, it was seen to be a serious problem. Attempts began to blossom to turn technological progress into an "endogenous" factor. Soon the qualifications of the laborer (human capital) began to be accepted as a significant factor in the technological process. But it had to be incorporated into growth models as an "endogenous" factor. It is fair to say that theoretical developments related to growth in the recent decades have been attempts to make both technological progress and human capital "endogenous" factors. According to Freeman-Soete, the "new growth theory", in fact, is nothing but the delayed incorporation of some realist assumptions into the neoclassical models that economic historians and Schumpeterian economists have known about for a long time. (Freeman-Soete;2003;341)

However, there is an interesting point which instantly attracts attention in the new growth models. Although everyone accepts that *"creative" mental labor* underlies technological innovations, the concepts like laborer, the qualification level of the laborer (human capital) and technology are evaluated as if they are completely independent and different from each other.

Separating laborer (L) from human capital (H) and, therefore, regarding or presenting them as being "different" factors is meaningless and causes misunderstanding. "New" endogenous growth theories give the impression that they are two separate and independent factors involved in production.

Since the existence of "completely unqualified" laborers is out of question, only we can speak only of laborers with "different" levels of qualification. In fact, there is no such thing as a "completely unqualified" laborer. Any attempt at introducing these two concepts as completely different concepts would only lead to an "artificial" and a "utopian" understanding of the true situation.

A "qualified" laborer that is endowed with both L and H produces new technologies by topping up their existing knowledge base. New technology is normally acquired through new empirical discovery and research. These new technologies are sometimes directed towards "lower unit cost" production or to an increase in quality. Sometimes new technologies create completely new products or methods of production. The origin of all these advances is the use of the laborer's *intellectual (mental) labor effort*. (Gürak;1993).

Consequently, the basic factor underlying the growth process is the *knowledge* produced and used by human beings. Since the source of new technologies is the "mental labor" of humans, we find that the most important source in long-run growth is a human being with a *"creative"* mental capacity. Neither capital nor anything else can replace or compensate for the contribution of *"creative mental labor"*. We should always keep in mind that;

Both the producer and the user of the knowledge in production is the "laborer".

In following sub-sections, we will first consider the neoclassical growth model followed by the growth models introduced since the 1950s.

Neoclassical Growth Theory-1: Pre-Solow

Before Solow's work I think it's safe to say that the dominant mainstream economic theorists (the neoclassical school) produced neither a viable nor a realistic long-run growth theory. The analyses they introduced were evaluated under the assumption that a state of full-employment 'equilibrium' existed. Since equilibrium implies a state of stasis i.e. nothing moves, it is natural to assume that it was only possible to describe growth until it reached the 'equilibrium' point. After having reached the 'equilibrium' state, there is only one cause of growth; population growth. Therefore, if population growth is regarded as an *exogenous* "given", when we add the *exogenously given* technology, then there remains only one source of economic growth until the equilibrium state is reached; i.e. investments. In fact, if we consider this from the neoclassical viewpoint it would be more accurate to describe this process as the attainment of the "state of natural equilibrium, rather than growth. However, it is possible to define this economic growth to the point of equilibrium as 'short-term growth', just like the 'short-term' growth that was described in Chapter-2, where the technology and the labor-force were a 'given'.

The concept of the 'quality of labor' or the 'qualifications of the laborer' is one of the most significant concepts ignored, to large extent, in analysis before Solow's work. In the equilibrium models, labor is mentioned as one of the two main production factors, but the differences in laborers' qualifications was not acknowledged. Labor was treated as a homogenous input in production, and therefore the concepts homogenous labor and homogenous capital seemed to be, accurate concepts in respect to the "logic" of the model; but the analyses were so far away from reflecting the reality.

Are there any "scientific" justifications for the neoclassical full-employment equilibrium model, as it appears to be so unrealistic and useless? Or, is it a question of the "accepted" ideology being "uncritically" accepted?

While looking for an answer to these questions, it is necessary thoroughly examine the time period during which the foundations of the neoclassical doctrine were laid down and to take into account the effects of this "ideology". In the second half of the 1800s, the influence of Marxist ideology was on the rise; Marx's claim that the 'workers have nothing to lose but their chains' was increasingly attracting sympathizers. This period was known as 'wild capitalism'. Marx's 'exploitation' doctrine became the capitalists' nightmare that was not into social reform or sharing their wealth. The environment was ripe for a proletarian revolution, and there were no alternative ideologies that were able to challenge Marx's revolutionary ideas.

As a result, Walras in Switzerland, Menger in Austria and Jevons in Britain retaliated, resulting in a proliferation of 'scientific' (?) economic models, which applied the so-called 'universal' laws. They did not pay any attention in their scientific (?) theories to any historical, cultural or psychological factors that could be involved in the formulation of a real economic model.

Interestingly, despite efforts to create a 'science' (like physics or astronomy); one of the basic concepts used was 'utility', a concept which has no place in physics or astronomy. Serious attempts were made to measure the magnitude of (cardinal) utility using mathematics. When they realized this was impossible they then shifted their focus from the concept of "cardinal" utility to "ordinal" utility, calling it a 'scientific' method.

When they thought they had found a permanent scientific solution to the differences in the quality of labor or the commodities by assuming 'homogeneity', they completely ignored the role of intellectual (mental) labor and technological innovation in their scientific models. The principle reason they ignored these factors was not due to a lack of skill of the scholars but due to their ideological stance. Thus, a social science known as 'political economy' was renamed 'economics' or 'pure economics', a scientific subject free from all values, feelings, history and the rest!

This scientific (?) approach since the 1870's has undergone many important changes. Economists today use incredibly advanced mathematical models and methods. But such models, unfortunately, are both extremely infertile and unrealistic. This is accepted not only by the opponents of the neoclassical ideology but by many reputable adherents of the Neoclassical school (see Gürak;2013).

Main features of the neoclassical equilibrium model before Solow are:

- Homogenous goods.
- The production function is homogenous at the 1st degree.
- Two homogenous production factors; labor and capital (K).
- K/L fixed.
- Price is a 'given'.
- $\pi = MPK*p$ and $w = MPL*p$ (Price and/or Profit is determined by the respective marginal efficiencies of the production factors).
- Full employment.
- The law of decreasing returns prevails (*DRTS*).
- Constant returns on scale (*CRTS*).
- Investment is determined by the amount of savings ($S = I = sY$).
- Say's Law prevails.
- Exogenous population growth ($\dot{L}/L = n$).
- In equilibrium, economic growth ends (*given n*):
- A 'given' technology.
- No technological innovations ($\Delta A/A = 0$).

The production function:

$$Y = F(K, L)$$

According to Figure: 4-1, 'given' the technology, as the employment of L increases, output would follow suit ($Y_2 > Y_1$); but according to the law of decreasing returns, as production increases, productivity per laborer is expected to decrease. D denotes the point of equilibrium.

Figure: 4-1 Production function

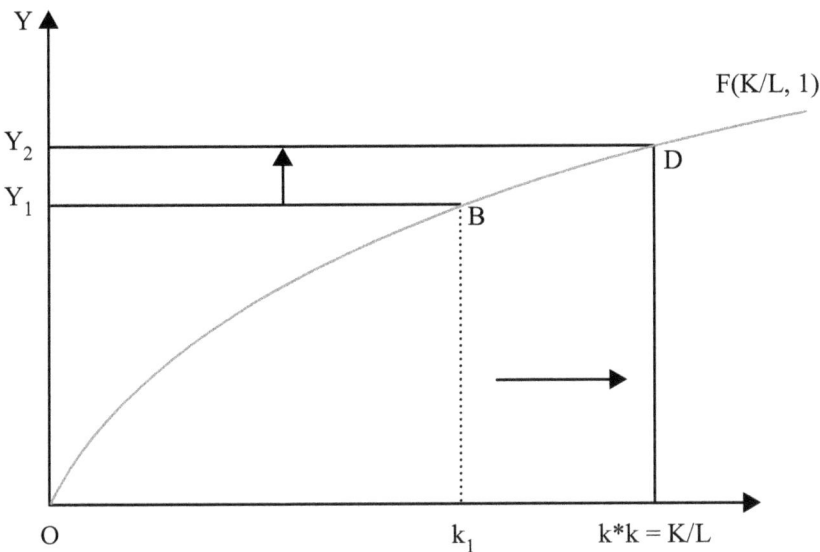

We will analyze growth in terms of output per worker (Y/L) in Figure: 4-2. Let's assume that there is unemployment at the beginning point B as there is not sufficient investment and the capital to labor ratio (K/L) is at the level of k_1. Y/L denotes the output to labor ratio, d the depreciation rate, n the population growth rate, $(d+n)k$ the amount of the investment required keeping k rate fixed, and sy denotes the savings per worker.

Actually, defining the movement of the economy from point k_1 to the equilibrium point of k^* as 'growth' seems incorrect or illogical. That is because, it is quite natural that at equilibrium, the economy is at point k^*. Any position of the economy other than at equilibrium would imply the inefficient use of resources. In other words, there is 'unemployment' at point k_1.

Figure: 4-2 Neoclassical growth and output per employee

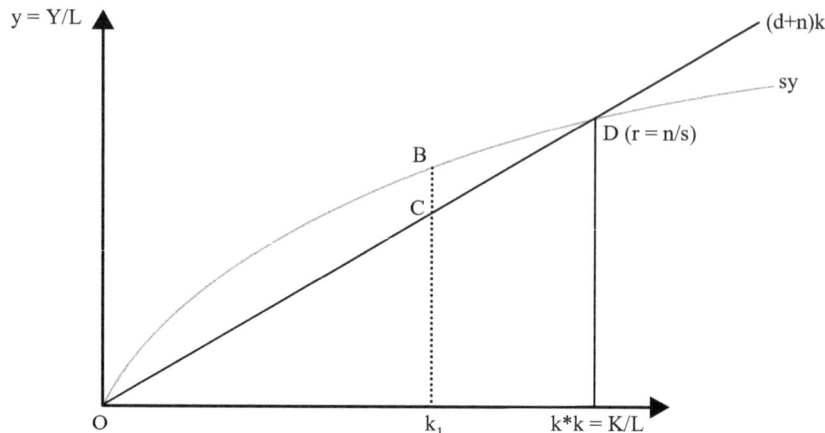

Since, at the point B, the capital per worker (k_1) is in excess of the investment or the savings (sy) by CD interval, namely $sy>(d+n)k$, higher levels of investment is required. In equilibrium, k and y are constant, while $r=n/s$. Thus, the growth rate (g) appears as a function of the saving rate (investment).

$$g = f(sY) = f(I) = f(\Delta K/L)$$

If we return to the model, to be in a stationary equilibrium, Y/L and K/L have to be constant. This is indicated by point of k^* in Figure: 4-3. D indicates the stationary equilibrium point according to the investment (saving) level, while D' indicates the equilibrium with regard to total output. Oy_1 denotes the amount of savings, while y_1y_2 denote the amount of consumption.

Figure: 4-3 Stationary equilibrium

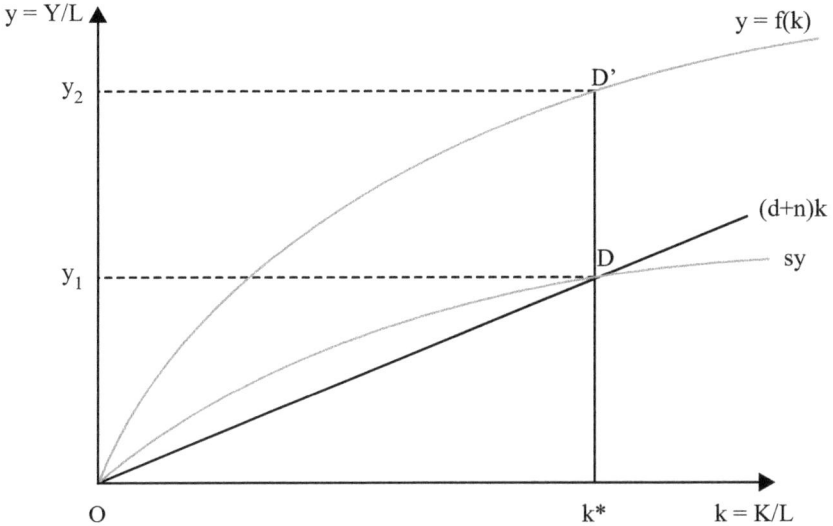

By assumption:

$f'(k) > 0$ ve $f''(k) < 0$

$Y=f(k)$ denotes the production function, (K/L) the capital-labor ratio, and (Y/L) the output to labor ratio. After having reached the stationary equilibrium level, the economic growth rate will be determined by population growth, 'given' the technology.

$g = n = s/v$

This theory of growth, which heavily influenced the economic thinking until the 1950s encountered a serious challenge as a result of the contribution of Solow whose model incorporated technological progress, while keeping the conventional 'equilibrium' state intact. Now, we will analyze this new approach more closely.

A Criticism of the Theory in Regard to Growth

- There is no attempt to establish a relationship between the institutional and the cultural infrastructure, in regard to the growth process.

- Both, the producers and the consumers are seen as "interactive robots" (homo economicus). Neither historical, nor cultural or humanitarian dynamics are considered.
- Because of equilibrium, there should not be an expectation of long-term "dynamic growth".
- No technological innovation.
- No relation between the quality of labor and the growth process.

Neoclassical Growth Theory-2: Solow & After

> "Please keep in mind that we are dealing with a drastically simplified story, a 'parable', which in my dictionary defines 'as a fictitious narrative or allegory'." (R. M. Solow, 1988)

The sole purpose of this section is to analyze the role played by the "technological change" entering into the economy from an unknown source, probably coming like "manna from heaven", thereby changing the "Total Factor Productivity" (TFP). Our focus will be on the concept of *"total factor productivity (TFP)"* due to *technical change"* meaning, in fact, *"technological change or progress"* (A). That is because; Solow's model (1957) introduces technological progress as the principal driving force of economic growth.

For Solow (1957), TFP was the key to long-run economic growth based on his research in the USA. Five years later in 1962, he had developed somewhat different ideas about the source of technological change and economic growth and assumed that "... *all technological progress needs to be "embodied" in newly produced capital goods before there can be any effect on output"*. He further assumed that "... *new technology can be introduced into the production process only through gross investment in new plant and equipment"*. (Solow;1962;76). But, the TFP idea presented in the 1957 article had already gained widespread acknowledgment around the world by researchers as the proper account of growth. Therefore, we shall focus our attention and criticism on the ideas presented in 1957.

TFP can be defined, given the factors of production (L and K), as the residual that causes increases in the total output. Or, alternatively, TFP is a variable which accounts for the increase in the total output not caused by the traditional inputs; labor and capital. The growth which cannot be explained by the increases in capital and labor input (the standard inputs of the neoclassical model), are explained as the *"Solow Residual"* or the *"Technology Residual"*. It is the measure of an economy's long-run technological progress.

Solow's study of 1957 which had *"re-discovered"* the crucial role of technological progress in growth process claimed that:

"Gross output per hour of work in the U.S. economy doubled between 1909 and 1949; and "some seven-eighths of that increase could be attributed to "technical change" in the broadest sense and only the remaining eighth could be attributed to a conventional increase in capital intensity." (Solow, 1987; xx).

But, there was a serious problem; according to Solow's model, the "technical progress" which was the driving force of growth was not due to endogenous factors inherent in the economic system. Rather, it was as if they originated from an unknown source with the help of "a magic wand". In this case, it would not be unfair to define growth as a "magical entity conjured up by a benevolent wizard".

In his own words, Solow's "distinguished three factors" of production are:

1. Straight labor.
2. Straight capital.
3. Technical change.

The basic assumptions of the model are:

1. Homogeneous output.
2. Homogeneous capital.
3. Homogeneous labor.
4. Factors are paid their marginal products.
5. Price is a "given".
6. Perfect competition.
7. Change in A is exogenous.
8. Initial production technology is a "given".

So the new production function containing the technological progress is:

$$Q = A f(K^{\alpha}, L^{\beta})$$

A has a "neutral" effect on both K and L. If the researcher chooses, the place of A may change and the function may appear as $Q = f(AK^{\alpha}, L^{\beta})$ or $Q = f(K^{\alpha}, AL^{\beta})$.

From the production function of Cobb-Douglas, Solow's growth function can be shown as:

$$\Delta Q/Q = \alpha\,(\Delta K/K) + \beta\,(\Delta L/L) + (\Delta A/A)$$

The change at A, $(\Delta A/A)$ is the total factor productivity (*TFP*) which can also be defined as the growth rate in output, given the quantities of capital and labor.

Estimation of TFP

Assume that the share of labor in the total income is 80 percent and grows at 1 percent while the share of capital is 20 percent and grows at 4 percent. Further assume that total income increases at 5 percent per annum. What is the rate of contribution of technological change per annum? Or what is the rate of TFP?

5 percent = (0.80*0.01) + (0.02*0.04) + TFP
TFP = 5 − 0.08 − 0.08 = 5 − 1.6
TFP = 3.4 percent.

In Figure: 4-4, growth is shown as a result of neutral technological innovation. As a result of technological innovation, the production per worker (q) increases and the new equilibrium point is D^{**}. While the amount of capital used per capita (k) remains fixed, the real wage increases (w_2-w_1). At the equilibrium state functional income distribution is as follows.

$w/r = MPP_L / MPP_K$

Figure: 4-4 Neutral technological progress and growth-1

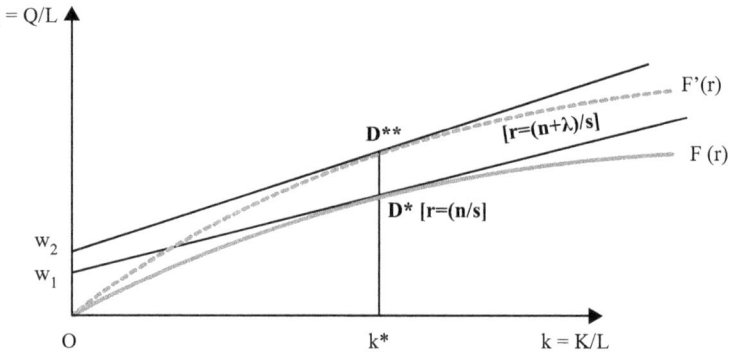

At equilibrium, the economic growth rate will be as much as the population growth rate (n) and the growth rate of the technological innovation (λ). Assuming that the population is constant, then the growth rate at D^{**} will be as much as λ. At the equilibrium k^* is constant; however the productivity increases due to technological innovation.

$g = n + \lambda$

We can demonstrate the effect of neutral technological innovation on the production from a different angle as in Figure: 4-5. Q shows the total output. As a result of technological innovation, the production curve shifts towards the right. Since k* is constant, output will increase as much as q_2-q_1.

Figure: 4-5 Neutral technological progress and growth-2

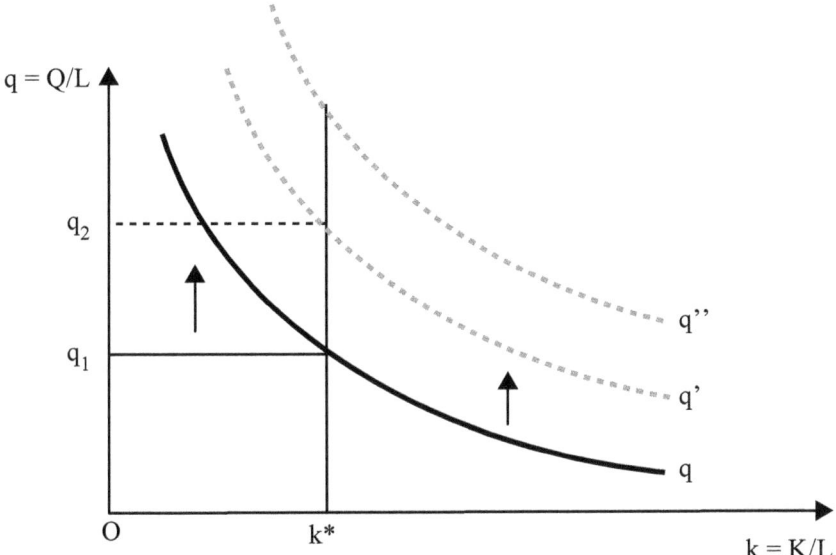

Criticism of Solow's Model & the TFP Approach

> "I described the aggregative theory of growth as a parable. You expect a parable to have a moral, but hardly to contain concrete instructions for the conduct of life." (R. M. Solow, 1988)

Solow's work substantially contributed to growth theory by "*re-introducing*" the importance of technological innovation in the process of growth. Nowadays, the role of technological innovation in the growth process has reached a stage where it cannot be ignored anymore. However, it is not realistic to assume the technological innovation is an "exogenous" factor, having an unknown origin. It seems that in order to save the "neoclassical equilibrium" analysis, Solow sacrificed the "embodiment idea" according to which: "... *much technological progress, maybe most of it, could find its way into actual production only with the use of new and different capital equipment.*" (Solow;1957;xxii). That is because; if the embodiment idea accepted, then capital equipment would no more be "homogenous". Solow kept the "homogeneity" assumption intact at the cost of a more realistic analysis.

However, despite its shortcomings the neoclassical growth theory has now an indispensable and non-ignorable production factor, *"technological progress"*.

TFP & Long-run Growth[13]

The sole purpose of this section is to analyze the role played by the "technological change" entering the economy from an unknown source, probably coming like "manna from heaven", thereby changing the "Total Factor Productivity" (TFP). For Solow (1957), TFP was the key to long-run economic growth based on his research in the USA. Five years later in 1962, he had changed his mind about the source of technological change from an "unknown" to a "known" source and assumed that *"… all technological progress needs to be "embodied" in newly produced capital goods before there can be any effect on output".* He further assumed that *"… new technology can be introduced into the production process only through gross investment in new plant and equipment".* (Solow;1962;76). But, the TFP idea presented in the 1957 article had already gained wide acknowledgment around the world by researchers as the proper account of growth. Therefore, we shall focus our attention and criticism on the ideas presented in 1957.

1- An area where the Solow's growth model fails seriously is in its ignoring of a *'services'* sector analysis; a sector which has the largest share of the income in all countries. That is because; the neoclassical models, even though they do not actually admit it, analyze only sectors which deal with producing physical goods. However, the relations of input-output in the services sector display different characteristics compared to the physical goods producing sectors. For example, measuring the growth in the entertainment sector or in the education sector by using the *TFP* method is a serious, if not impossible task. Or how can the *TFP* measure the tourism sector when the tourist season changes from five months to eight months? The ignoring of the services sector alone is sufficient to reject or dismiss Solow's model as a long-run growth model capable of explaining real economies.

2- Solow's growth analysis is based on the assumption that the commodities produced are *'homogenous',* i.e. there is only one type of commodity produced. This circumstance, inevitably, leads us the conclusion that: Technological innovation only contributes to produce a *"given"* product, with a *'new method of production'* as described in Chapter-2. In other words, the new technology

---

13 In order to properly evaluate the ideas, critics and comments in this section the reader is advised to refresh her/his knowledge on the ideas, definitions and comments mentioned in the Chapter-2 of this book.

aimed to reduce the unit production cost of the "homogenous" product. Otherwise, the capital-owner will have no incentive to introduce the new technology.

3- The new technology, according to Solow's model, is a capital-saving technology according to our definitions in Chapter-2 of this book. With the employment of new technology, new "homogenous" capital-goods are now employed for a longer period in supplying the homogenous goods, thus reducing the unit fixed cost of supply. In Solow's words:

> "... if the capital stock consisted of a million identical machines and if each one as it wore out was replaced by a more durable machine of the same annual capacity, the stock of capital as measured would surely increase." (Solow;1957;314).

4- Following Solow's example of "a million identical machines", a technological change occurs through a magic wand. After the technological change, the annual supply of the "identical" machines (Q= 1,000,000 pieces) and the quantity of "identical" capital-goods (K= 1,000,000 pieces) do not change. Q and K are "the same" goods in equal quantities regarding the "homogeneity" assumption. What is the implication of technological change if there is no quantitative change? The difference is, say, the capital-goods were employed one-year before they wore out and replaced before the technological change, and now they are more durable and, say, wear out after two years. With the employment of the new "magic wand" technology, the unit cost of production is now lower. And, by assumption, "... *the stock of capital as measured would surely increase.*" (Solow;1957;314).

5- By the way, there are two rather critical details in Solow's model which should not be overlooked. The first is that the "homogenous good" or the "identical machines" of the model is not only a capital-good, but also a raw-material input, an intermediary-good and a consumer-good, at the same time. The second critical detail is that the "homogenous" good is one single unit "block". It is a machine without any demountable parts. In other words, the machine contains no screws, no separable units, no bottoms, etc. Otherwise, the "homogeneity" assumption would not be fulfilled. What a perfect and miraculous model; is it not?

Let's give another example; assume that the commodity produced is a car. Accordingly, now and in the future, only "identical" cars shall be produced. Technological progress can only be used for the purpose of increasing the quantity of identical cars thus reducing per unit fixed cost.

Question-1: Assume that our society consists of one million persons and that each person owns a car. Before the technological progress, we used to replace the worn-out cars every year with a new one. After the technological progress, the quantity of the cars supplied has not changed; it is still one million cars per year. But now, the cars are more durable and, say, the depreciation period of the cars increased from one year to two years. Would the technological progress not affect the supply-demand equilibrium adversely? Would a technological progress become a source of instability?

Question-2: Let's assume that one year after the "magical" appearance of technological progress, a newer technology is introduced. Further assume that the new technology increases the durability of the cars from two to four years. After the first technological progress, the life of our cars, i.e., the depreciation period, had increased and we have used the new cars for one year only. That means the present cars can be used for another year. But, the second technological progress makes the cars even more durable; four years instead of two. What would be the rational behavior? Start using the new technology immediately? Or wait for a year until the cars wear out?

Question-3: According to Solow's model, a technological progress includes improvements in the labor-force quality. If the origin of improved quality in the labor-force is "endogenous", then how do we know what the properties of the new technology are in order that we can provide the appropriate education/training for the labor force?

In fact, the assumption that technological progress is "exogenous" and supposedly of "unknown" origin, according to Solow, is of itself sufficient to refute Solow's Residual approach to growth. *Technological progress is developed by creative mental labor for a certain purpose* and offered to the markets as being *incorporated* into the physical product (capital goods, intermediate goods and end-products). In other words, technological change, which is supposed to be exogenous and of an unknown origin is the product of labor and serves to enhance the productivity of per unit labor-time employed or to introduce new products to the market.

Another objection to the *TFP* approach is in conjunction with the new technology being a "public good". Thus, by assumption, every producer can easily access the new technology and employ it in production without any restriction. If some countries, e.g. the developing countries, cannot make use of them efficiently, the reason is either insufficient investments or an inadequacy on the part of the labor force (human capital). Or both may prevail. Because, problems arising from rapid population growth along with the many other problems of the developing countries inhibits the quantity of the qualified laborers available.

Technological innovation is, except for "blackboard economics", not a public good. Free access to technology is restricted by patent laws, both in developed and developing countries. However, *the practices in relation to the ownership of technology* appear to be an important impediment to the efficient use of technology and of growth in the developing countries. In addition, the scarcity of qualified laborers and an inadequate technological infrastructure only compounds their problems. The *ownership and the control of technology* play a very important role in the development of any country.

Solow's model is criticized also for failing to explain the differences in the development levels of different countries and why the factor prices fail to converge as a result of foreign trade or why do some countries "acquire more advanced technology" as a heavenly gift than the others? The developed countries seem to be more favored regarding the acquisition of technological change. I wonder why?

We have defined technology in economic terms very briefly as the "*knowledge necessary for production*" (see Chapter-1 of this book). Technology, (or 'knowledge'), is incorporated into products which are essentially, natural resources that have been transformed by the human labor. In other words, all man-made products incorporate some form of technology produced by the physical and intellectual ingenuity of humans. Therefore, it would be meaningless and seriously mistaken to try to demonstrate that capital goods (K) and laborers (L) are separate from the technology (A) discussed in TFP analyses. It would be like saying "gravity does not exist because I can't see it".

How is it then a clever and skilled person like Solow failed to establish a relationship between technological progress and "capital goods" without incorporating the notion of creative intelligence? Was it his blind adherence to an ideology that prevented him from researching and finding the source of technological progress?

In fact, Solow did establish such a relationship. According to his broad definition of technological development, the "improvements" in the labor-force quality are included in the technological progress (Solow, 1957; 312; & 1988; xix). In addition, Solow thought that his model did not sufficiently reflect the importance of investment and claimed that a substantial part of technological progress is "*embodied*" or "*materialized*" in the products supplied. In Solow's words, most of technological progress;"… *could find its way into actual production only with the use of new and different capital equipment.*" (Solow;1957;xxii)". But later he preferred to put aside the "embodiment" idea. That is because; Denison (1988), a highly respected colleague by Solow, thinks that his "embodiment" idea is not sufficiently descriptive (1988; XXIII).

In summary; in addition to the well-known shortages or inadequacies of the neoclassical models, the weakest side of Solow's model is that he *fails to account for the source of the technological progress and the introduction of "new" products.*

For a continuous long-term growth, given the natural resources, there has to be technological innovation which introduces 'new goods and/or new production methods" in addition to innovations which lead to 'new production methods for "given" products'. In the absence of "new" products the market would saturate after a while and the long-run growth would depend on exogenous miracles. It is uncertain that Solow's model can even explain short-term growth.

Some TFP-related Data

By taking a look at the Table: 4-1, we observe that during 1950-1999, the GDP of the USA increased at 3.6 percent annually, of which 1.2 percent was due to capital accumulation, 1.3 percent due to the increase in the number of laborers, while the remaining 1.1 percent was due to technological development.

Table: 4-1 Sources of growth in the USA

	Growth rate $\Delta Y/Y =$	Capital $\alpha \Delta K/K +$	Labor-force $\beta \Delta L/L +$	TFP $\Delta A/A$
	(annual average growth)			
1950-1999	**3.6**	**1.2**	**1.3**	**1.1**
1950-1960	3.3	1.0	1.0	1.3
1960-1970	4.4	1.4	1.2	1.8
1970-1980	3.6	1.4	1.2	1.0
1980-1990	3.4	1.2	1.6	0.6
1990-1999	3.7	1.2	1.6	0.9

Source: US Department of Commerce, US Department of Labor, **in** G. N. Mankiw (2003), Macroeconomics; 233, Table: 8-3.

The question is; why those "magical" progressive changes in technology do not take place in the developing countries in the first place?

According to some research which measures the growth rate in south Korea on the basis of Solow's model (Pyo;2001), the growth generated by technological progress between 1946 and 1999 was zero (see Table: 4-2). However, not only researchers, but the ordinary man on the street knows that this does not reflect the true state of affairs. In order to see the extent of technological progress in South Korea, using a rough guide, just take a look at the automotive or cell-phone industry there in the last 20 years.

Table: 4-2 Annual average growth rates in S. Korea between 1946-1999

Total Factor Productivity	0.0
Added value	6.6
Labor inputs	3.7
Capital input	10.4
Factor inputs- Total	6.6

Source: H. K. Pyo (2001) Economic Growth in Korea (1911-1999); 98 Table: 23, *Seoul Journal of Economics,* Vol. 14, No: 1

In contrast to results of Pyo, the results of another researcher in South Korea showed that, TFP is calculated to increase on average 1.5 percent annually between 1960 and 1994. (Table: 4-3).

Table: 4-3 Annual average growth rates in S. Korea between 1960-1994

	Per Capita	Contribution to	Output Growth
Output growth rate per worker	Growth rate of physical capital	Education	Total Factor Productivity
5.7	3.3	0.8	1.5

Source: Collins & Bosworth (1996); in D. Rodrik, (1999), Yeni Küresel Ekonomi ve Gelişmekte Olan Ülkeler 51; Table-3.1; Sabah Kitapları, Istanbul.

Considering both calculations are performed in accordance with the *TFP* criteria, which one is wrong?
One of them?
Or both?

TFP and Growth in Turkey

Here are some interesting data regarding *TFP*. The Turkish National Productivity Center (MPM) suggests in its Productivity Report published for the year 2004 that *'growth in Turkey is not based on productivity growth'*. It reaches this conclusion by referring to various researches of the State Planning Organization (SPO), MPM and some independent economists (MPM, 2005, 25). According to a SPO publication, while the contribution by *TFP* to growth in Turkey was 9.5 percent between the years 1972 and 1991, it was - (minus) 2.1 percent between 1992 and 2000. Again according to a Productivity Report, during the 1990s, the contribution of the *TFP* to growth was 100.1 percent in Sweden where it seems to have had a remarkable, even an extraordinary performance. Whereas in Japan it was minus (-) 52.6 percent in the same time period (see Table: 4-4).

Who would give credit to such a kind of high-level "scientific" (!) analysis and conclusions?

Table: 4-4 Estimated contribution of TFP to growth in some countries (1990s)

Country	(%)
Turkey	-2.1
Japan	-52.6
Sweden	100.1

Source: OECD: in; Verimlilik Raporu-3, 2004, 26, MPM, Ankara

According to the *TFP* approach, during the 1990s in Turkey and particularly in Japan, the growth process went backwards. Especially in Japan, the productivity issue appeared to be *VERY SERIOUS (?)*. Because, the contribution of the *TFP* for growth was *minus 52.6 percent* (Table: 4-4). It looks like technological innovation contributed negatively to the growth of Japan during the 1990s. Probably, the Japanese are not aware of the seriousness (?) of their situation. Yet, here we are, while we were all thinking that the Japanese cared so much about technological innovation and we even thought that if they weren't the first, they were definitely the second best in this area.

Well according to the *TFP, how mistaken we were!*

The data in Table: 4-5 shows clearly the level of importance that the Japanese place on R&D, thus on technological innovation.

Table: 4-5 Per capita R&D expenditure (1997-1998) (in US dollar)

Japan	858.4
USA	465.9
Turkey	4.8

Source: Lall-Albaladejo (2002), "Indicators of the Relative Importance of IPRs in Developing Countries: *in*; Verimlilik Raporu-3, 2004, 49, Table: 12 MPM, Ankara

Could there be a problem" with the TFP method?

"Endogenous" Growth Models

Since 1980s, new growth theories have been developed. These propose an "endogenous" model of growth and assert that growth comes from "within". These models place their emphasis on the quality of "human capital" (labor quality) and that technological advance is the result of this "human capital" (labor). Actually these "endogenous" growth theories are not all that new. Their origins are found in the works of Classical economic scholars such as; A. Smith, Ricardo, particularly Marx; and latterly, Schumpeter[14], among others. These "new" endogenous growth models only found a common ground in the economy of a "hypothetical world", not in the real economies that we live in. A theory which is unable to form a relationship between growth, technological progress and qualified labor (human capital) is trapped in the vicious circle of hypothetical models.

In the following sub-section, we shall evaluate the endogenous growth models developed by Lucas, P. Romer, Institutional School, Aghion-Howitt, Grossman-Helpman and Mankiw. In addition, Eichengreen's approach to the "end of growth" will be evaluated.

Lucas: the Mechanical Model of Growth

In 1988, Lucas tried to explain growth with a model built upon human capital as an endogenous factor. However, as Lucas admitted, the mechanics of economic development model that he introduced reflects an "artificial" world composed of "interactive robots". Everything was so computable and measurable that his model, he claimed, could be tested by computers at any time. Lucas ascribes the term "mechanical" to his model but, nevertheless, he asserts in his defense, that such a model is capable of reflecting the real situation. His purpose is to construct: "… a neoclassical theory of growth and international trade that is consistent with some of the main features of economic development." (Lucas;1988;3).

Lucas considers "growth" and "development" as two distinct fields. According to Lucas, human capital which is acquired through education and/or learning-by-doing, which enhances the qualifications of the laborer, should be considered as an alternative to technological progress or at least considered as a "complementary engine of the growth".

Lucas' Models

Lucas develops three models in order to explain *"growth and international trade"* within the framework of the neoclassical doctrine.

1. The first model emphasizes *physical capital accumulation* and *technological change*.
2. The second model emphasizes *human capital accumulation* through *schooling*.
3. In the third model, he emphasizes *specialized human capital accumulation* through *learning-by-doing"*.

---

14 We will evaluate in detail the theories of some Classical economists as well as some selected "contemporary" views in the following Chapters.

General Assumptions

1. Population growth is exogenous and a given.
2. All exchange is barter, i.e., goods-for-goods.

Model-1

Assumptions:

1- Single output.
2- A closed economy.
3- Competitive markets.
3. Identical rational agents.
4. Constant Returns to Technology.
5. At date t there are $N(t)$ persons/man-hours devoted to production.

The first model which is an application of a neoclassical model closely following Solow's and Denison's works emphasizes *physical capital accumulation* and *technological change*. Lucas argues that; such neoclassical growth models attach unnecessary importance to technological innovation and ignore other important factors. Above all, these models ignore the contribution of human capital. In other words, Lucas considers that Solow's and Denison's works are "not satisfactory" and claims that they were: "... *attempting to account for the main features of U.S. economic growth, not to provide a theory of economic development.*" (Lucas;1988;7). And Lucas continues his criticism: "*By assigning so great a role to 'technology' as a source of growth, the theory is obliged to assign correspondingly minor roles to everything else.*" (Lucas;1988;15).

Regarding the contribution of human capital in the development process as significant, Lucas notes that:

> "*Human knowledge is just human, not Japanese or Chinese or Korean. I think when we talk in this way about differences in 'technology' across countries we are not talking about 'knowledge' in general, but about the knowledge of particular people, or perhaps particular subcultures of people.*" (Lucas;1988;15).

The most important aspect is the knowledge used to increase productivity. He wants;

> "*... a formalism that leads us to think about individual decisions to acquire knowledge, and about the consequences of these decisions for productivity. The body of theory that does this is called the theory of 'human capital'.*" (Lucas;1988;15).

Lucas introduces two central reasons to account why the neoclassical growth theory does not stand for a useful theory of economic development:

> "... its apparent inability to account for observed diversity across countries and its strong and evidently counterfactual prediction that international trade should induce rapid movement toward equality in capital-labor ratios and factor prices." (Lucas;1988;17).

As an alternative to the first model, Lucas presents two further models founded on the concept of "human capital acquisition".

Model-2a: Human Capital (Through Schooling)
In the second model, Lucas aims at presenting "an alternative" or at least "a complementary" driving force to technological innovation. For this purpose, first of all, he injects the *"human capital"* factor into the model of Solow. Thus, human capital constitutes the basis of the second and third models. As specified previously, the concept of human capital simply implies the *"general skill level"* $h(t)$.

> "The theory of human capital focuses on the fact that the way an individual allocates his time over various activities in the current period affects his productivity, or his $h(t)$ level, in future periods." (Lucas;1988;17).

How the individual uses his/her time impacts on the current and future acquisition of human capital, therefore, the productivity of the laborer. Lucas calls it "aspects of technology".

The assumptions of this model:

1. Technological level is a "given".
2. Technological innovations are excluded.
3. There is no technological gap between nations.
4. A *"single good"* is produced.
5. There are two types of capital:

    a) *Physical capital* which has the conventional features as in the neoclassical models:
    b) *Human capital* which increases the productivity of both the laborer and the physical capital.

6. The level of the acquisition of human capital of all individuals is the same.

Lucas assumes that the laborer spares $u(t)$ of his/her time for production, and the remaining $1-u(t)$ for the acquisition of human capital. Accordingly, a laborer having a human capital level of $h(t)$ is as productive as two laborers having a human capital level of $1/2\ h(t)$ or two workers having an $h(t)$ level of human capital but working for only half a day. There are a total of N laborers who have h skill level. The Production Function is:

$$Y = F(K, N^e)$$

where, N^e denotes the quantity of effective laborers. The growth rate of human capital, $\hat{h}(t)$, also determines the development rate of the economy.

$$\hat{h}(t) = h(t)\, \delta[1-u(t)]$$

where, $u(t)$ is time given to production, $1-u(t)$ given to education. while δ is a parameter which indicates the growth rate of human capital.

According to the equation above;

> "... if no effort is devoted to human capital accumulation, [$u(t)=1$], then none accumulates. If all effort is devoted to this purpose [$u(t)=0$], $h(t)$ grows at its maximal rate δ. In between these extremes, there are no diminishing returns to the stock $h(t)$." (Lucas;1988;19).

In other words, when $u(t)=0$, since the workers give their time entirely to education, then human capital acquisition is achieved at its maximum level. But in this case, no production takes place. According to this perspective, *the time spared for education is the time taken away from production*. If individuals want education, then they have to postpone some of their "current" work time. As a result their "consumption" is either partly or totally postponed. Then it follows that if the acquisition of human capital is to be increased by means of education, then individuals need to be subsidized.

> "Economies that are initially poor will remain poor, relatively, though their long-run rate of income growth will be the same as that of initially (and permanently) wealthier economies. A world of consisting of such economies, then, each operating autarchically, would exhibit uniform rates of growth across countries and would maintain a perfectly stable distribution of income and wealth over time." (Lucas;1988;39).

According to Lucas, this new model is in accordance with the neoclassical models of Solow and Denison which have already become standard and are founded on technological innovation as well as the developments of the US economy during the 19th Century. However, though the model seems:

> "... capable of accounting for average rates of growth, it contains no forces to account for diversity over countries or over time within a country." (Lucas;1988;40).

What conclusions can be derived from the studies of Model-2a? The reply from Lucas is:

> "Normatively, it seems to me, very little." (1988:27).

So, according to Lucas, this new model is not capable of explaining the dynamic growth process over time, neither on global nor national basis.

In his quest to find a solution to the differences in global growth rates and income levels, Lucas could not arrive at a satisfactory explanation, which caused him to head for another model consisting of the production of two commodities.

Model-2b: "Learning-by-doing" and Comparative Advantage

In the third model, Lucas tries to develop a *"specialized human capital acquisition"* model taking a different approach to the standard neoclassical growth model. He highlights the concepts of *"learning-by-doing"* or *"on-the-job-training"* and *"specialized human capital"* which are different from schooling but are at least as important as schooling. The new model assumes that;

> *"... human capital accumulation is taken to be specific to the production of particular goods, and is acquired on-the-job-training or through learning-by-doing. If different goods are taken to have different potentials for human capital growth, then the same considerations of comparative advantage that determine which goods get produced where will also dictate each country's rate of human capital growth."* (Lucas;1988;40).

The assumptions of the model:

1. Technology is a "given".
2. Preferences are "given".
3. Closed economy.
4. There is no physical capital.
5. There are two consumer goods; C_1 and C_2.
6. Relative prices are determined by the human capital endowments.
7. The population is constant.
8. Since *productivity growth* originating from human capital is *infinite*, the law of increasing returns prevails.
9. Human capital acquisition is *"exogenous"*.

The production function of the consumer goods "*i*";

$$C_i(t) = h_i(t)\, u_i(t)\, N(t) \quad i = 1,2$$

where, $h_i(t)$ is the *human capital specialized* to the production of good *i*, and $u_i(t)$ is the fraction of the laborers devoted to the production of good *i*. So;

$$u_i \geq 0 \text{ and } u_1 + u_2 = 1$$

The growth of human capital (learning-by-doing) $h_i(t)$ is related to the quantity of laborers $u_i(t)$ devoted to the production of good *i*. In other words, the amount of laborers allocated to the production of a good will lead to an increase in the total human capital by virtue of learning-by-doing.

If we denote the skill-ability acquisition rate of a certain amount of the labor force (u_i) with growth rate of δ_i, the growth of "specialized" human capital will be as follows;

$$\hat{h}_i(t) = h_i(t)\, \delta_i\, u_i(t)$$

The acquisition of human capital in both sectors depends on the average level of skill in the industry. In other words, if we assume that δ_1 is a product of a more advanced technology, then it will be;

$$\delta_1 > \delta_2$$

Namely, in a country where goods are produced with more advanced technology, human capital acquisition will grow at a higher rate. At the beginning, δ is at higher rate for each product, but this rate gradually declines over the course of time due to diminishing returns. As in the preceding model; "... *human capital will lose its status as an engine of growth.*" (Lucas;1988;28).

Lucas wants equation $\hat{h}_i(t) = h_i(t)\, \delta_i\, u_i(t)$:

> "... *to 'stand for', then, is an environment in which new goods are continually being introduced, with diminishing returns to learning on each of them separately, and with human capital specialized to old goods being 'inherited' in some way by new goods.*" (Lucas;1988;28).

As "new" products are constantly introduced into the market, so a new process of learning-by-doing begins. Old skills are "somehow" transferred into new products. Otherwise, the rate of human capital acquisition and the growth rate might one day end. Therefore, the ongoing human capital acquisition is sustained thanks to *"new products"*.

The assumption of exogenous human capital acquisition through learning-by-doing is problematic. In Lucas' words;

> "*As was the case with the human capital model of the preceding section, it is obvious that the equilibrium paths we have just calculated will not be efficient. Since learning effects are assumed to be external, agents do not take them into account. If they did, they would allocate labor toward the 'high δ_i' good, ... so as to take advantage of its higher growth potential.*" (Lucas;1988;31)

The agents will not take an exogenous factor into account and will not be shifted towards the higher technology fields. In this case, both the growth rate of the unspecialized labor force and the growth rate of the overall economy will be lower than it should. To avoid such incidences, governments have to promote the production of high technology goods.

Introducing "International Trade" into the Model

According to the assumptions above, international trade was not taken into account, namely, a "closed" economy was assumed. The inclusion of foreign trade will cause significant changes in the model. The assumptions of the model:

1. Perfectly free trade.
2. Two final goods.
3. A continuum of small countries.
4. Prices are a "given" and equal for all countries at a "world price" level (1, p).

Figure: 4-6;

> "... gives a snapshot of this world at a single point of time. The contour lines in this figure are intended to depict a joint distribution of countries by their initial human capital endowments. A country is a point (h_1, h_2), and the distribution indicates the concentration of countries at various endowment levels.
>
> At a given world price p, countries above the indicated line are producers of good 2, since for them $h_1/h_2 < p$ and they maximize the value of their production by specializing in this good. Countries below the line specialize in producing good 1, for the same reason... Clearly, the supply of good 2 is an increasing function of p and of good 1 a decreasing function." (Lucas;1988;31).

Countries remaining above the OR price line will be specialized in the production of the second good. Their h_1 endowments remain fixed while their h_1 endowments grow at the rate δ_2. For those countries remaining below axis OR, they will specialize in the production of good 1 so that its h_2 is constant while h_1 grows at the rate δ_2. Thus the distribution of endowments determines the goods supplied over time. As a result, globally, the maximum amount of value will be produced.

Figure: 4-6 Lucas' growth model and the global distribution of labor

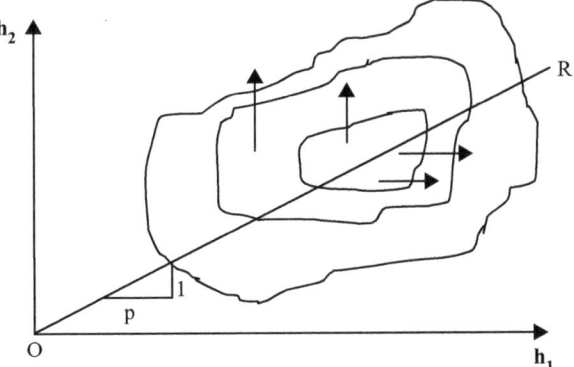

> "In this set-up, human capital accumulation is taken to be specific to the production of particular goods, and is acquired on-the-job-training or through learning-by-doing. If different goods are taken to have different potentials for human capital growth, then the same considerations of comparative advantage that determine which goods get produced where will also dictate each country's rate of human capital growth" (Lucas;1988;40).

To summarize; Lucas believes that his third model is capable of explaining *"wide and sustained differences in growth rates across countries"*. However Lucas confesses that: *"With a fixed set of goods, which was the only case I considered, this account of cross-country differences does not leave room for within-country changes in growth rates."* (Lucas;1988;41)

Conclusions drawn by Lucas:

1. Countries which start out poor will remain relatively poor.
2. Long-term growth rate will be the *"same"* in all the countries.
3. Long-term income and welfare distribution will be "perfectly" stable.
4. "If in capital goods are introduced into this model world economy, with labor assumed immobile, there will be no tendency to trade." (1988; 39)

Criticism of Lucas' Model

A critical area where the Lucas' model seems to fail is in the *'services'* sector; a sector which has the largest share of the income in all countries. Because, the neoclassical models, even though they do not actually admit it, analyze only sectors which deal with producing physical goods. Any theory or model of growth or development failing to explain the service sector's activities is bound to be sterile and inefficient. Ignorance of the services sector alone is sufficient to reject or dismiss Lucas' model as a long-run growth model capable of explaining the real economies.

Lucas' model, in fact, does not have much to say about the real world. It describes a "mechanical" world and "mechanical" relations between robots in an "artificial" world. Such fictitious models can be interesting for "blackboard" economics satisfying only for the neoclassical economists. There is very little real and useful information, as is the case for all neoclassical doctrines. However, although the model is mechanical, the robots and the settings "artificial", Lucas' study on demonstrating the importance of human capital, which we define as "qualified labor", in the growth process, indicates an important development. What is emphasized is that "human capital" is the engine which drives growth and an increase in the value of production is in line with the acquisition of human capital. The higher the level of skills and knowledge of the laborers, the higher the rate of growth will be. However, Lucas is aware of the difficulty in observing and directly measuring of the quantity of human capital (h) or its rate of growth (δ).

Lucas' critical error is that he fails to establish a relationship among "intellectual labor", technological innovation and products. There are "new goods" in his model, but no "technological progress" in a realistic sense. Since technology is a "given", "new" goods are not produced by virtue of new technologies. Then, where do "new goods" come from? Since formal schooling does not exist and human capital acquisition starts growing by virtue of the learning-by-doing process subsequent to the production of "new goods", the origin of these new goods has to be the human capital. Lucas' model fails to build a bridge between human labor and the creation of technological progress as presented in Chapter-2 of this book.

Despite of all of its shortcomings, according to Lucas' model, human capital, that is the qualified laborer, is the sole input and also the reason for growth. In this case, we have a right to ask this question: What is the essential difference, between Marx who claims that the origin of growth is the *labor-power* of the proletariat and Lucas who defends that the origin of growth is *human capital (qualified laborer)*? From both perspectives; the main factor producing growth is the *"laborer"*. Isn't it?

Then, can we say that Lucas is a *"disguised"* Neoclassical Marxist?

The Institutional School

According to the supporters of school of institutional economics, the founder of this movement of thought is T. Veblen (1857-1929), although he did not define himself as an institutional economist. J.R. Commons and W.C. Mitchell are acknowledged as other important economists who are identified with this school. Among other renowned economists are G. Myrdal, R. Coase, C.E. Ayres, A. Gruchy, D.C. North and G. Hodgson.

What is an "Institution"?
"Institution" can be defined as various rules and restrictions which are established by humans and aim at maximizing the benefits which can be classified under two categories:

1- Formal institutions (i.e. laws, ownership rights, etc.).
2- Informal institutions (traditions, ideologies, behavioral patterns etc.)

If we use the game of football as an analogy; a football game has certain formal rules. For example, the game is played by 11 players, one of which is a goalkeeper. Offences committed in the penalty area are punished with a penalty kick.

Namely, misconduct has a price. Spectators watching the game in the stands react to the decisions according to informal behavioral patterns developed by the traditions and customs of their community. And by cheering for his/her team, spectators try to destroy the concentration of the opposing team.

Economies, too, are governed and directed by formal and informal rules. Formal rules are normally expressed in writing. Informal rules, on the other hand, though normally not written and explicit, have always been present and have always had an impact on the system just like the formal rules. Especially in the communities with a deeply traditional structure, informal rules may sometimes be more influential and decisive than the formal rules. For example, bribery is formally prohibited in every country. However, it is considered "a matter of course" in many countries. It is possible to change the formal rules with changes in legal regulations in a short time. But to change the informal rules generally requires a much longer period of time.

In an economic community, institutions have to support their productive activities. An "institutional framework" influenced by culture and ideologies can sometimes clear the way for economic growth, at other times it is detrimental to the expectations of a particular country. For example, the general expectation in the US is that regardless of the type of reform, no-one thinks that such reforms will damage private ownership rights and the right to do business. This enhances the confidence in the system as well as its effectiveness. Since such a tradition has not developed to the same extent in Russia and countries of former Soviet Bloc, despite radical changes to the system during the 1980s. Consequently, confidence and the effectiveness level of the system were very low especially during the first decade following the 1980 changes and therefore many economic problems occurred.

"Institutional Economics"

In answer to the question: What is *institutional economics* about, Özveren provides the following answer:

> "The institutional economics departs from the emphasis that the processes in the traditional field of economics are, in fact, institutions, by their nature...... In other words, elements such as market and firm in the main field of mainstream economics are each, in fact, institutions, although mainstream economics resists recognizing them as such. Since no economics can be thought of without them, they are institutions by the very definition of economics. Thus, economics can be defined and understood as an institutionalized process." (Özveren, 2007; 16-17; translated by the author)

By this definition, institutions cannot be considered in isolation from the economic processes. With the advent of an economic process, economic institutions begin to emerge and, then change and develop according to the conditions influencing them over a period of time. Therefore, the science of economics or any economic analyses naturally includes economic institutions. The "pure economics" system of the neoclassical doctrine, which is purified from the notion of institutions, does not reflect reality. Thus, it is requested by many economists, especially heterodox ones, which "economics" should be restored to its rightful place as a "social science" and includes all other related sciences like politics, sociological behavior, psychology, etc.

We have noted that institutional economics analyzes and evaluates the economic process together with its institutions. In this respect, we can point out that the Classical economists such as Smith and Marx were in fact "institutional economists". Classical economists may not have developed comprehensive theories and analyses like the institutional economists of today, but we cannot say that they were unaware of, or indifferent to, the existence, the importance and the effects of the institutions on the economy. Therefore, it would not be a mistake to suggest that "institutional economics" was initiated by the Classical economists. The advocates of the "German Historical School" can also be recognized as "institutional economists" in this context. In fact, each non-Neoclassical economist can be regarded as an "institutional" economist to some respect.

As far as we are aware, there has been no formal "institutional" growth theory developed by the economists defining themselves as "institutional economists". However, we know some institutional economists who expressed opinions about economic growth. The following section is a close analysis of the opinions about growth of D.C. North, and R.J. Barro, respectively, two of today's widely respected economists.

D.C. North on Growth

According to North who is originally an economic historian, the restrictions established by people shapes the interactions among people (North;2002;p.9). In other words, institutions are the *rules of the game* for ordinary people. However, we should avoid confusing institutions and organizations. Organizations are formed by groups of people such as political parties, companies, unions, sport clubs, and other associations. Organizations also shape the interaction between people; but they are different from "institutions". The main goal of institutions is to define and determine rules, whilst organizations aim at "reaching certain goals".

Despite not proposing a theory of growth, D.C. North is a researcher who is curious, like many other economists, about why countries are at different welfare levels and why they grow at different rates. North criticizes the neoclassical theories for not having an historical dimension that he considers necessary to understand the issue of welfare differences. Because, suggests North, "yesterday determines the choices of today and tomorrow" (North;2002;Preface). Therefore, the notion of "institution", which is not incorporated into the Neoclassic economic theories, are, for North, very important in terms of their formation, development, and impact on the community. The most significant factor, for him, regarding long-term economic growth, is the remedies or the barriers introduced by institutions on complex economic relationships. Although some criteria used for achieving long-run growth are useful, others do not provide the benefits expected. He suggests, one of the main reasons for such problems is the lack of the *appropriate* social and political *institutions* which are expected to make structural economic reforms successful. Institutional frameworks and institutional changes which are useful in decreasing "uncertainty" are critical.

North cited the developments that occurred in the US during the 19th Century as an example of the contribution of institutional changes on economic performance. He points out that as a result of "appropriate" institutional changes economic growth increased and education was encouraged. He shows that the US government managed to realize institutional changes required in time.

North provided further examples of the long time value of institutions by using European historical perspectives from the Middle Ages up to the 1900s. Facts such as the promotion of the right of private ownership, the weakening of the conservative guild system accelerated the industrialization of countries such as England and the Netherlands. However, the industrialization process in Spain where institutional changes were slow and private ownership rights were underdeveloped economic progress did not move forward.

After having analyzed historical developments in the developed countries, North concluded that the most important problem for the developing countries was their "institutional structure". North notes that opportunities for entrepreneurs are still complex (North;2002;17). He claims that existing institutions encourage monopolism, instead of competition; they restrict new opportunities, instead of creating or developing new opportunities. For example, he suggests the importance of education is still not sufficiently understood in many developing countries. Let's assume that two countries having same level of development make the same economic reforms at the same time, but the implementation of these reforms are different. The country where reforms are implemented properly is likely to have higher chance to succeed.

For example, rulers may use the resources of their country in order to keep themselves in power, instead of using their country's resources to increase production which would be in step with the accepted objective criteria. This kind of behavior can be seen in many of the developing countries. In such cases, the national economy will sustain damage. As a result of the waste or misuse of resources for the sake of personal or political gain these particular countries head the world corruption league, while their economy struggles.

To summarize, according to North; effective institutions are key to the growth.

An Evaluation with a Critical Eye

It is not possible not to agree "in principle" with the conclusions drawn by North about the contribution of the institutional structure on economic development. However, when we take the current global economic and political conditions into account, it would be "naive" to think that a "proper" change in the institutional structure of a developing country would be sufficient to enhance its economic performance and enable it to develop to the level of the developed OECD countries. That is because; the "efficient institutions" of the developed countries have arrived at their present situations to the detriment of "other" countries.

For example, in our age, the *imperfections (defects) in the global technological markets and the global production relationships are the major barriers to sustained "long-run" growth in developing countries. The prevailing global imperfections reduce their ability over a given period of time, to attain to the same levels of welfare found in the more developed countries (convergence concept)*. The good and "efficient" institutions in the West have to change first, in order to pave the way to make the developing country institutions "efficient".

Brain-drain is another area of significance regarding the "efficiency" of institutions. No doubt the Western economies have a lot to gain from the brain-drain. Yet every "efficient" move in this direction from the Western institutions is to the detriment of developing country economies. Developing countries need educated brains much more than the developed countries do. But do the Western "efficient" institutions care about that?

Without changing the content, the scope or the conditions of the present global institutional structure and global relationships, the aim of closing the gap between the developed and the developing countries may only be possible in a "fictitious" world. For a country's stable long-term growth in a global context, all "institutional" defects created by the developed countries' "efficient" institutions must be eliminated from all market.

P. Romer: An Endogenous Growth Theory

Human Capital, R&D & Technological Progress

After Solow's claim that around 80 percent of growth in the US originates from technological progress, the importance given to technological progress in the analyses of growth has increased significantly since the 1950s. However, in Solow's model technological progress was an exogenous factor. Another significant development was that studies concerning the qualification level of the laborer (human capital) have significantly increased since the 1960s. Consequently, the number of growth models which emphasized technological progress and human capital acquisition has significantly increased.

In fact, the seeds of these ideas concerning technological progress and the labor-force qualification level had been shown long time ago in the period of Classical economists such as Smith, Ricardo and Marx. Since the 1870s, with the efforts to transform "political economics" into a "positive science" like the natural sciences, technology was, in general, regarded as a "given" and the laborers as a homogeneous input in the production relations. The role of technological progress which was emphasized by Marx had almost been forgotten until economists like Schumpeter and then Solow resurrected it. However, in those days, the models which establish *a direct and organic relationship between the qualification levels of the laborer (human capital), technological innovation (new products and/or new methods of productions) and the growth process* were still not thought of.

In such circumstances, the contribution of P. Romer (1990) added a new dimension to the theory of growth. As righteously criticized by Romer (1994), it is impossible to gain an insight into long-run growth process by just accumulating homogeneous capital goods. That is because, when the markets consisting of homogenous goods reach a saturation point after a certain length of time, the "static equilibrium" of neoclassical doctrine is reached, but the growth process comes to an end. Yet, as the real economic conditions prove, the growth process, thus, the enrichment of countries is a dynamic and cyclically changing process; never a "static" equilibrium.

With the contributions of Romer, the neoclassical growth theory with its *"stationary equilibrium"* based on the law of diminishing returns took another major blow. According to the new theory, as a result of employing the *new designs (new technologies)* which are the products of human capital and their introduction into the production function as a productive input, *the law of increasing returns (IRTS)* would prevail. "Endogenous" technological innovations which were known during the period of Classical economists but disappeared after the Marginalist Revolution, which were ignored by the all "hallowed" static equilibrium models, now began once again to take their place on the economic stage. *"New"* technologies, now, are not manna from heaven, but are *"intentionally"* added to the economy as a consequence of internal dynamics. Consequently, the concept of "static equilibrium" is dismissed by Romer; but revived as *equilibrium growth*.

Romer's efforts to understand long-run growth and his attempts to make technological innovation endogenous, was a significant contribution to the theory of growth. Romer suggests appropriate economic policies which will lay the foundations for an institutional framework supporting the progress of technology for a sustainable growth process. He says:

> "For a nation as a whole, an effective institutional arrangement for supporting technological advance must ... support a high level of exploration and research in both private firms and in universities."(Romer;1994).

The Basic Assumptions of Romer's Growth Model
Romer's growth model is based on three premises:

1- *"Technological change"* lies at the heart of economic growth.
2- The model is one of *"endogenous"* technological change which arises from intentional investment decisions made by profit- maximizing firms.
3- "New technology" is a *"non-rival"* good. Once the cost of creating a new set of knowledge has been incurred, this new knowledge can be used over and over again at no additional cost. In other words, once it is produced, the marginal cost of the re-utilization of a new technology is zero. Its use by someone will not decrease the amount available for the use of others.

However, the use of non-rival goods (the new technology) by others can be excluded by the legal system. *"A good is excludable if the owner can prevent others from using it."* (Romer;1990;74). This is called "partial excludability" and implies only a partial access to the new designs (new technologies).

Human capital (H), as "an *exogenous*" factor, is a "rival good". Use of H by someone excludes others from using it and therefore it is excludable. That is because, the human capital is a specific faculty of human beings and the person holding the human capital cannot be present in multiple places and perform jobs simultaneously.

Four Basic Inputs of the Model:

1. Capital goods (K) are measured in units of consumption goods.
2. Labor services (L) are skills such as eye-hand coordination, measured by counts of people.
3. Human capital (H) measured in years of formal education and on-the-job training that are person specific.
4. The level of technology (A) is measured in numbers of designs.

The Other Simplifying Assumptions of the Model:

1. Population (N) and the supply of labor (L) are constant.
2. Total stock of human capital (H) in the population and its fraction supplied to the market is constant.
3. Intermediate goods sector (X) and the final goods sector (Y) use the *SAME* technology.
4. Only knowledge and H are used to produce new designs or knowledge. L and K are not used at all.

The Role of Knowledge in the Growth Process

According to Romer, "knowledge" contributes in the growth process in two different ways:

1. *New designs* (new technologies) (ΔA).
2. *New (additional) knowledge* (ΔB) improves the efficiency of H in R&D.

A restriction may be imposed on the use of A in the supply of K. But there is no restriction on using ΔB (additional knowledge) in the R&D process. Therefore, there is a potential for everyone to make use of this *knowledge* (ΔB) in R&D, which is an idealized and a non-rival good, in order to create "new" designs, thus further increase growth.

The assumption is that knowledge is essentially a non-rival good, enables us to draw some logical conclusions on the knowledge spill-over effect. Thanks to spill-over effects of knowledge, it becomes an infinite growth potential. By virtue of the "productivity" of knowledge, the *"increasing returns"* (IRTS) prevails on the economy as a whole, instead of the conventional constant returns to scale" (CRTS).

Analysis of the Growth Model

The model consists of three sectors:

Sectors

a. *The research sector;* produces new designs (A).
b. *The intermediate-goods sector (X).*
c. *The final-goods sector (Y).*

a- The Research Sector:

1. New designs (technologies) are produced and patented by making use of only the given knowledge stocks (B) and the human capital (H_A), making use of various combinations of the raw materials.
2. L and K are not used at all in the *Research* sector.

3. Technologies (*A*) and "knowledge" are *"idealized goods"*. They are not materialized or internalized in the physical goods.[15]
4. New designs (*A*) and "knowledge" (*B*) which are non-rival goods may grow without bound.
5. *A* can be measured easily, because *A* is the count of the number of designs.
6. The measurement of H_A is calculated by the years of formal education and on-the-job training provided to individuals.
7. Everyone can freely study and learn from the stock of knowledge (*B*) and designs (*A*) but is not allowed to produce. Instead they can produce new designs by utilizing the existing knowledge.
8. Each unit of new knowledge corresponds to a design for a new good.
9. The law of increasing returns to scale (*IRTS*) prevails in the sector.
10. Free entry and exit from the sector and perfect competition.
11. The patent of a new technology is sold to only *one Monopolist producer* in the intermediate-sector for the supply of "capital goods".
12. Since the market is open to competition, the patent revenue that is raised will be equivalent to the present net value of the revenue of the "monopolist" in the intermediate goods sector.
13. Due to perfect competition, the profit will be zero.
14. Larger markets, not larger populations, spark off higher *Research* and growth.

The growth rate in A:

$$\dot{A} = \delta H_A A$$

where $\dot{A}$ indicates the growth new technologies, H_A the total human capital employed in research and δ the productivity parameter of the human capital (the capability of creating new designs). The equation above has two substantive and two functional assumptions (Romer;1990;83).

Substantive assumptions:

1. Devoting more human capital to research leads to a higher rate of supply of new technologies. The higher H_A, means the higher *A*.
2. The larger the stock of technologies and knowledge is, the higher the productivity of a H_A in the research sector.

---

[15] According to this assumption, technology (or design) is nothing but *"pure KNOWLEDGE"*.

Functional form assumptions: The amount of designs is linear in each of H_A and A when the other is held constant. Linearity in HA is not important for the dynamic properties of the model. However, linearity in A is what makes unbounded growth possible (Romer;1990;84).

b- *The Intermediate Goods (X_i) Sector:*

1. Sector inputs are K, L_y and H_y saved in the final-goods sector.
2. Marginal productivity of H_Y increases at the same rate as A in the *Research* sector.
3. Firms buy infinitely lived patents of new designs (A) produced in the *Research* sector in order to produce (X_i) *units of durable capital-goods.* Since durables do not depreciate, they may be used forever.
4. Since the patent right is owned by one single firm, this firm is a MONOPOLY and each firm produces "*only one capital-good*".
5. Every capital-goods produced is different from the others; therefore they are not homogeneous.
6. New capital-goods are not a substitute of the former ones.
7. The K produced is *rented* to the final-goods sector. The value of one unit of durable "i" is the present discounted value of the infinite stream of rental income that it generates.
8. Monopolist's rent ($p=AC+\pi$) is the same for every firm in the sector.
9. Constant returns to scale (*CRTS*).

Growth of the intermediate goods sector depends on the level of consumption. The higher the savings from consumption, the higher the production of capital goods will be. $K(t)$ represents the total stock of capital goods and $C(t)$ aggregate consumption at time t. $K(t)$ evolves according to the rule:

$$K(t) = Y(t) - C(t)$$

The production function of K:

$$K = \eta \sum_{i=1}^{a} X_i = \eta \sum_{i=1}^{A} X_i$$

η denotes the quantity of forgone consumption and used in the intermediate goods sector; while *i* denotes the quantity of capital goods produced. "*Thus, H and L are fixed, and K grows by the amount of forgone consumption.*" (Romer;1990;82).

c- Final-goods Sector (Y):

1. The inputs of the sector are K, H_y and L_Y which are the *"rented"* capital goods from intermediate-sector.
2. Output is either consumed or used to produce additional capital goods. That is to say, the foregone consumption of final-goods is used in the intermediate-goods sector to produce capital-goods.
3. Only physical goods are produced.
4. Perfect competition.
5. Price is a given.
6. The law of constant returns to scale (*CRTS*) prevails in the sector.
7. The technology used is the *SAME* as in the intermediate goods sector.
8. The use of capital goods increases until the profit rate is equal to the marginal product of capital ($r=MPP^K$).
9. Since entry into the market is free in the *Research* and final-goods sectors, wages will equalize in both sectors ($w_a=w_y$).

Final-goods sector firms earn zero profits and own no assets. (Romer;1990;88). In the sector where capital goods can perfectly substitute each other; and the homogeneous first degree Cobb-Douglas production function for output is:

$$Y(H_y, L, x) = H_y^\alpha L^\beta \sum_{i=1}^\alpha x_i^{1-\alpha-\beta}$$

Balanced Growth

According to Romer's model, while *perfect competition exists in Research and final-goods markets, monopolistic competition* prevails in the *intermediate-sector* and a *"balanced growth"* prevails.

> *"An equilibrium for this model will be paths for prices and quantities such that (i) consumers make savings and consumption decisions taking interest rates as given; (ii) holders of human capital decide whether to work in the research sector or the manufacturing sector taking as given the stock of total knowledge A, the price of designs P_A, and the wage rate in the manufacturing sector w_A; (iii) final-goods producers choose labor, human capital, and a list of differentiated durables taken prices as given; (iv) each firm that owns a design and manufactures a producer durable maximizes profit taking as given the interest rate and the downward-sloping demand curve it faces, and setting prices to maximize profits; (v) firms contemplating entry into business of producing a durable take prices for designs as given; and the supply of each good is equal to the demand."* (Romer;1990;88).

While the equilibrium is at point E at the beginning, it will shift to E^* as a result of growth (see Figure: 4-7). The condition determining the allocation of H between the research sector and final-goods sector says that the wages paid to H in each sector have to be equal. The most suitable environment for growth would be the lack of any difference between the private and social returns. If this cannot be sustained, then the second best policy would be to support the total human capital acquisition.

Figure: 4-7 Equilibrium growth

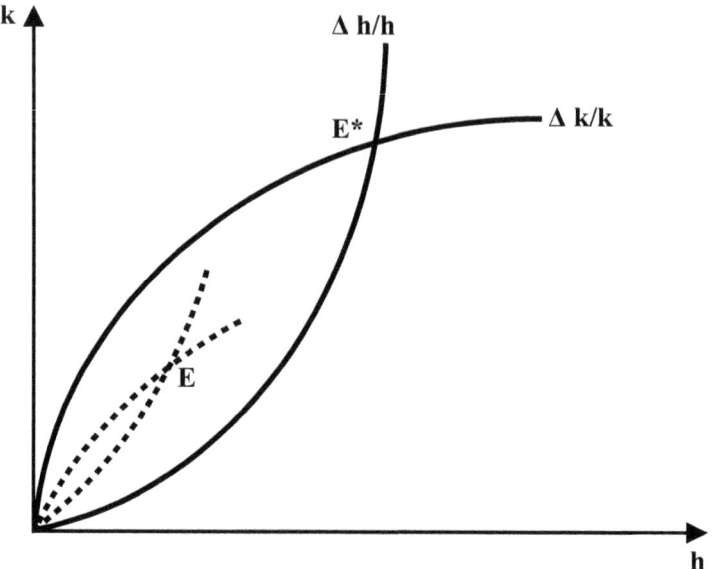

To summarize Romer's conclusions;

1- The engine of growth is the Research sector consisting of the stock of knowledge (B) and human capital (H).
2- The research sector produces new designs (new technologies).
3- All the research is embodied in capital-goods.
4- Each new unit of A corresponds to a design for a "new capital-good".
5- The rate of technological change is sensitive to the rate of interest.
6- The larger the human capital accumulation is, the higher the growth rate will be. That is because; "new designs" will be introduced into the market along with human capital acquisition and this will increase the growth rate.
7- Free international trade can act to speed up growth.
8- Low levels of human capital may help to explain why growth is not observed in underdeveloped economies that are closed.

9- A developing economy with a large population can benefit from economic integration with the rest of the world.

A Critical Evaluation

Romer's model built upon "human capital" (qualified labor ") and "new designs or technologies" contributed significantly to mainstream growth theory and produced some novel ideas. For instance, in the model technological innovation, as the engine of long-term growth, becomes endogenous. Furthermore the model emphasized that the origin new technology is knowledge produced by human capital. That's to say, long-run growth is realized due to new knowledge, which is the product of the mental labor. Thus, the *new designs or technologies, the products of mental labor are the source of long-run growth*. Romer does not explicitly state these points but they are the logical conclusions to this model.

Despite all of its valuable contributions, Romer's growth model can be criticized in many ways.

Some Criticisms:

1. Romer's growth model is founded on the production of "physical goods". Therefore, it does not have a word to say concerning the developments about the output and employment in *service* sector, which has the largest share in GDPs across the world. A theory which does not explain the growth process in the service/ sector cannot be claimed to be a "general" theory of long-run growth. This failure alone is sufficient to reject Romer's model outright.
2. A design (A) is an "idealized" knowledge and produced without any physical tools and/or equipment (K) or corporeal labor (L). This is a simplifying assumption of the model but makes the model look over-simplified. This is possible in a "utopian" economy only, i.e., in the minds of the neoclassical economists.
3. Romer uses the concepts *human capital (H)* and *labor (L)* as if they are two different production factors. This is a serious mistake. In fact, they are substantively the same thing: the *laborer.* If not ideological, then what is the reason of such an artificial dichotomy? The purpose may be to distinguish the labor "without skills or qualifications" from the "skilled or qualified" ones. But this would not be a satisfactory explanation. Surely, a more rational division of "laborer" would be as "qualified" and "non-qualified". Or "skilled" and "non-skilled". By the way, a completely unqualified or unskilled laborer does not exist in any sector or in any country. Every laborer possesses some degree of qualification, some more, some less. Romer seems to be aware of the shortcomings of such an artificial dichotomy, but he ignores the situation in order to keep his model intact.

4. The model presents the research-sector, where new technologies are developed, as an independent profit unit separate from the intermediate-sector. In addition, both firms in respective sectors possess "monopoly" rights over the new technology, though the research sector firm sells it to the latter sector. This is far from reflecting the reality of the situation. In fact, usually, the inventor of the new technology and the owner of it with monopolistic right to use it are the same.
5. According to Romer's model, B and A have the potential of infinite growth. One may have the impression that all producers have equal opportunities in accessing knowledge and growth; but the actual situation is different. Although patent knowledge on documents about the new technologies is accessible by everyone, not every producer has the same opportunities of using of them to the level and extent they would like to. Patented knowledge may act as a source of inspiration for "other" producers, but only to the extent of the adaptive and absorption capabilities of the firm which, in turn is limited by its technological development level and the qualification level of the laborers it has employed. Firms in developed countries, have in general, such a knowledge base, a qualified labor-force and a technological infrastructure to benefit from patented knowledge. Since the technological difference between them is not substantial, it would be relatively easier for developed country firms to imitate the new technology and introduce similar products. But, in general, the issue is completely different for the companies of the developing countries which lack many significant inputs. For example, there is:
   - An adequate general level of knowledge.
   - The amount of the qualified laborers and their qualifications for R&D is, in general, inadequate.
   - The infrastructure for the advance technology related R&D is inadequate.
   - Funds required for R&D are insufficient.
   - There are many serious global market imperfections for the transfer of technology.
   - Ever increasing global investments and the global distribution of labor causes the producers in developing countries to specialize in the production of labor-intensive products which require relatively less advanced technologies.
6. The neglect of the impact of *"institutional and cultural"* frameworks; a serious shortcoming of the mainstream theories is no different in Romer's model.
7. Romer also neglects to consider the *"historical"* perspective. How countries have arrived at their present conditions, is uncertain. What kind of measurements was taken to promote growth in the past? Was there any protection

of the economy? Did crisis and cyclical fluctuations occur? If they occurred, then how the countries concerned did overcome them? What are the economic implications of imperialistic and colonialist policies? These questions are not asked and the appropriate answers are not sought.

8. According to Romer's assumption, *"new" designs do not substitute the former ones.* Then, the opportunity of producing the existing products at a lower unit cost with new methods of production disappears and this is a very serious drawback in explaining the real developments that occur. That is because; a substantial part of technological innovation is concerned with producing the "existing" products at lower costs. By introducing cost-reducing technology, the producer gains an important cost advantage over the competitors.

9. Capital goods do not depreciate; they can be used forever. This assumption is another one of the highly utopic assumptions of Romer's model. Just like the assumption that no labor is employed in the intermediate sector.

10. One of the important shortcomings of the model is its incompetence in explaining the differences in global growth rates. According to Romer, the larger the human capital accumulation is, the higher the growth rate will be. However, the model is incompetent in explaining why countries such as Russia or Ukraine that are relatively richer in human capital produce less "new" designs than the countries such as the US or Japan.

11. The model is not applicable across the world. That is because; the developing countries' economies may sustain economic growth for a long period of time by simply *transferring* the "known technologies" from the developed countries without engaging in the costly *R&D* process. In other words, developing countries are not required to produce "new designs" in order to grow. What the developing countries "urgently" require is *qualified laborers who have the ability to use the transferred technologies effectively*, instead of a *creative minded labor force* producing "new" designs. Of course, assuming that there are no technology market imperfections.

12. *"Uncertainty"* is one of the most significant and a realistic feature of the real world, but unfortunately it does not have a place in Romer's model. If uncertainty did not exist and caused risk, everyone having access to some funds would be a potential entrepreneur. Furthermore, the problems such as cyclical fluctuations and underemployment would not occur.

13. We observe in Romer's model many unrealistic assumptions such as "equilibrium", "a given price", "perfect knowledge", "perfect competition" all of which belong to the frequently referred to, "utopian" world of the

119

neoclassical economists. This shows clearly that it is not easy to give up ideological obsessions.

14. Romer, who is so sensitive about the "new technology and human capital", can be seriously criticized for not beginning his analysis with *a value/price theory which embraces "technological innovations" and "creative mental labor"*. Value/price theory is the foundation stone of all other theories. In light of this, we should modify Romer's question[16] and ask:

Where is the discussion of innovation, invention, discovery and technical progress in the "value-price" theory?

If we are to make a concise comment about the Romer's growth model, it can be said that:

It is an important contribution in respect of his efforts to demonstrate the relationships of the qualification level of the laborer (H), knowledge (B) and technological innovations (A) to long-term growth. Yet, in order to understand "real" economic relations in a realistic fashion, it seems to be inadequate just as all neoclassical models.

A final note: As in Lucas' model, Romer's model too presents the *"qualification level of the laborer"* (human capital) as a key concept in creating growth as well as being the origin of productivity enhancements. Then, do you think that Romer's approach is essentially different from Marx's approach?

Could Romer be defined as a "disguised" Neoclassical Marxist?

R.J. Barro on Growth

R.J. Barro is a renowned economist who has produced several works on the subject of "endogenous growth". Barro produced in 1990 an article which presented a model which concerned a discussion on "public expenditures" which we will discuss first. Then we will discuss the basic factors involving the "convergence" of national economies, as presented in his work, published in 1999.

Public Expenditures and Growth - 1990

In his article in 1990, Barro develops a simple endogenous growth model with two sectors based on constant returns (CRTS) assumption "to a broad concept of capital". Barro wanted to show how public expenditure financed by taxes affected growth or benefitted the national economy. According to his model, saving and growth rates may be at suboptimal levels due to public expenditure and

---

16 Romer's question: "Where is the discussion of innovation, invention, discovery and technical progress?"

externalities related to taxes. Therefore, there are interesting options between public policies on one side, and saving and growth rates, on the other.

According to Barro, as the beneficial productive government spending increases, the saving and growth rates rise at the beginning, but then decline. "*With an income tax, the decentralized choices of growth and saving are 'too low', but if the production function is Cobb-Douglas, the optimizing government still satisfies a natural condition for productive efficiency.*" (Barro;1990;103). According to Barro: "*Empirical evidence across countries supports some of the hypothesis about government and growth.*" (Barro;1990;103).

In concluding this section of his article, he draws three major conclusions relating to the long-term development of the U.S.: (Barro;1990;148).

1. "*Public policies can exert a quantitatively large influence on the average growth rates of economies operating in isolation. Policies can display these effects because they influence private incentives for accumulation of physical and human capital.... In both open and closed economies, relatively small changes in tax rates can lead countries to stagnate or even regress for lengthy periods, if these policies eliminate incentives for growth.*
2. *The effects of taxation depend importantly on aspects of production technology for new human capital, about which there is presently insufficient information. In part, this reflects the fact that our human capital good is a composite of many different activities and that we have not taken a sufficiently precise stand on its essential content. On the other hand, there has been little research in labor economics... on the parameters of individual technologies for investment in human capital.*
3. *... since policies have the potential to influence the growth rate in models with endogenous long-run growth, there is generally much larger quantitative influence of policies on welfare than in the neoclassical model.*"

Barro concludes that: "*... incentive effects of policy can influence economic activity: taxation can readily lead to development traps and growth miracles.*" (Barro;1990;148).

A Critical Evaluation

It would be a serious error to claim that the main or one of the main causes of long-run economic growth is public policies. That is because; the appropriate public policies can only help to create the appropriate infrastructure for the development and utilization of technological progress. In other words, appropriate public policies will contribute indirectly to "long-run" growth, as long as they are supported by, and allow creative skills to develop to create new technologies.

The Determinants of Growth - 1999

Burro, in his book published in 1999, after having identified the basic determinants of growth, investigated whether economies at different levels of development could close the gap and converge at point in time.

He proposed that the basic determinants of the growth are:
1. High levels of education;
2. Health conditions, especially life expectancy;
3. Lower rates of fertility;
4. Lower rates of government transfers;
5. The legal system;
6. Advantageous terms of trade.

> "If all economies were intrinsically the same except for their starting capital intensities, then convergence would apply in an absolute sense, that is, poor places would tend to grow faster per capita than rich ones." (Barro;1999;1).

The concept of capital covers both physical goods and human capital in the form of education, experience and health. Economies have very many different characteristics for the convergence. Some of them are:

a) Saving tendencies.
b) The number of children per family.
c) The willingness to work.
d) The level of technological development.
e) The policies that are implemented by the government.

Under the circumstances where such gaps exist, the convergence of economies can only occur under "certain specified conditions". Because, due to the reasons listed above, the stationary-equilibrium level of the capital to labor ratio (K/L) and the output to labor ratio (Y/L) would be different. Barro states, that it is possible to add the "imperfections" in the private ownership rights together with local and foreign market imperfections to the list of the differences outlined above.

According to Barro's definition, the concept of capital does not consist only of capital-goods. Education, experience and health levels that are included in "human capital" are also included in the concept of capital as a whole. Therefore, capital (K) includes both physical capital goods and human capital.

As per the extended neoclassical model, growth rate of per capita output is defined as:

$$D_y = f(y, y^*)$$

where D_y denotes the growth rate of per capita output; y denotes the current level of per capita output; while y^* denotes the long-run level of per capita output.

"... if the government improves the climate for business activity –say, by reducing burdens from regulation, corruption, and taxation or by enhancing property rights- the growth rate increases for a while." (Barro;1999;9).

So the conclusions drawn by Barro are as follows:

"With respect to government policies, the evidence indicates that the growth rate of real per capita GDP is enhanced by better maintenance of the rule of law, smaller government consumption and lower inflation. Increases in political rights initially increase growth but tend to retard growth once a moderate level of democracy has been attained. Growth is also stimulated by greater starting levels of life expectancy, and of male secondary and higher schooling, lower fertility rates, and improvements in the terms of trade." (Barro;1999;119).

For given values of these variables, the country with the lower GDP per capita (the poorer one) will grow faster. In other words, these findings point out the likelihood of a *"conditional convergence (an intersection at a point)"* of different economies (Barro;1999;119).

Additional public policies that are likely to be important for growth are (Barro;1999;120):

1. Tax distortions.
2. Pension and other transfer payments.
3. Regulations that affect labor, financial and other markets.
4. Infrastructure investments.
5- R&D outlays.
6. The quality of education.
7. Income and wealth distribution.

A Critical Evaluation

Barro builds his arguments on the fictitious assumptions of "equilibrium", which is a traditional but fruitless obsession of the neoclassical doctrine. However, he has a significant difference when he reintroduces the idea that political decisions are influential in determining the economic prosperity of a country: A subject that the neoclassical scholars had either forgotten or chosen to ignore since the 1870's.

Otherwise, Barro's ideas are not related to a long-run growth theory which is our main topic of interest. He only makes some attempt to describe the influential aspects of public spending on the growth process.

Aghion-Howitt: Creative Destruction

Aghion-Howitt published an article "A Growth Model through Creative Destruction" in the periodical Econometrica in 1992 inspired by Schumpeter's ideas on technological change and recognized it as "endogenous". The model was constructed on the "equilibrium" assumption by following the mainstream tradition of economics. According to the authors, the source of growth is the *"vertical technological innovations"* taking place in the competitive research sector. The vertical technological innovations improved the quality of the products. The new and better products render the previous ones obsolete by virtue of the innovation. This process is called as the *"growth through creative destruction"*. By assumption, growth results exclusively from the innovations.

In Chapter 2 of this book, two types of technological innovations were introduced: 1- *"new products"*; and/or 2- *"new production methods"* as distinct from the efficiency growth process with a given technology. A "new product" was, usually, accompanied with "new production method" (process technology). The second type of analysis, *"new production methods"*, however, referred to a "cost-saving" process innovation of a "given" product. (The vertical technological innovation in Aghion-Howitt's model, corresponds to our first type of innovation; a "new product", not to a cost-saving process innovation.

Particulars of the Model

1- The expected growth rate of the economy depends on the technological innovations in research sector.
2- Technological innovations are "endogenous".
3- The reason behind technological innovations is the competition in the research sector.
4- Every innovation introduces "new" intermediate goods. With the application of such innovations consumer goods are produced more efficiently.
5- A patent is obtained for the resulting innovation, thus the firm realizes a monopolistic profit. The driving factors pushing firms towards research and inventions are these monopolistic profits.
6- In the next period, after innovation, the new and better products render the former products obsolete. A new monopolistic period starts, due to the new innovation.
7- Employment is constant in the research sector.
8- *GDP* grows at a random rate.

Effects of the Research in a Given Period:

The amount of research performed in a given period depends negatively upon the expected amount next period, through two effects on the economy:

1. "The first effect is that of creative destruction. The payoff from research this period[17] is the prospect of monopoly rents next period. Those rents will last only until the next innovation occurs, at which time the knowledge underlying the rents will be rendered obsolete. Thus the expected present value of the rents depends negatively upon the Poisson arrival rate of the next innovation. The expectation of more research next period will increase that arrival rate, and hence will discourage research this period.
2. The second effect is a general equilibrium effect working through the wage of skilled labor, which can be used either in research or in manufacturing. [...] the expectation of more research in the next period must correspond to an expectation of higher demand for skilled labor in research next period, which implies the expectation of a higher real wage. Higher wages next period will reduce the monopoly rent. [...] Thus the expectation of more research next period will discourage research this period by reducing the flow of rents expected."

The functional relationship between researches in two successive periods has a unique fixed point, which defines a stationary equilibrium. The stationary equilibrium exhibits balanced growth, in the sense that the allocation of skilled labor between research and manufacturing remains unchanged." (Aghion-Howitt;1992;324).

A Basic Growth Model
Assumptions:

1. There are three tradable objects:
   a) Labor;
   b) One consumption good;
   c) One intermediate good.
2. Infinitely-lived individuals with identical preferences defined over life-time consumption.
3. Constant rate of time preference rate is constant; $r > 0$.
4. The marginal utility of consumption is constant.
5. r is also the interest rate.

Categories of Labor:

There are three categories of labor:

1. *Unskilled labor* (M): employed in the production of *consumer goods*.
2. *Skilled labor* (N): employed either in research or in the *intermediate*-sectors.
3. *Specialized labor* (R): employed only in *research*.

Each individual is endowed with a one-unit flow of labor. A labor is either unskilled or skilled or specialized.

---

17 A period: the time between two successive innovations.

Sectors and Production

There are three sectors in the model.

The Research Sector:

- Competitive firms.
- Employs the skilled labor (N) and the specialized labor (R)
- Introduces technological innovations at random.
- An innovation means a "new" invention to be employed in the intermediate goods sector. As a result of the employment of the innovation in the intermediate goods sector, consumer goods can be produced more efficiently.
- Specialized labor (R) is essential for the sector for growth. Otherwise no technological innovation takes place.

The Intermediate-goods (x) Sector:

- Only skilled labor (N) is employed.
- Innovation coming from research sector is used in the production to produce the intermediate goods more efficiently.
- The producer is a monopolist due to his/her owning a patent right and earns monopolist rent.

The Consumer Goods Sector:

- Only employs unskilled labor (M) and the inputs produced in the intermediate sector.
- Number of skilled labor is constant.
- Constant returns to scale ($CRTS$).

Production function:

$$y = AF(x) \qquad F' > 0, F'' < 0$$

where, y denotes the output; x the intermediate goods; while A denotes the *productivity parameter* of the intermediate goods. Employing the "*new*" intermediate goods in the production of consumer goods increases the productivity parameter by the amount of $\gamma > 1$.

$$A_t = A_0 \gamma^t \qquad (t = 0, 1, \ldots\ldots)$$

where, A_0, denotes initial value of the development process.

Regarding the implications of new technology, Aghion-Howitt gives the following example:

"Real-world examples include such 'input' innovations as the steam engine, the airplane, and the computer, whose use made possible new methods of production in mining, transportation, and banking, with economy-wide effects." (Aghion-Howitt;1992;328).

Evaluation of the Basic Model:

1- The source of growth is the vertical technological innovations in research sector.
2- Technological innovation exists and is endogenous.
3- New technologies are produced by the *specialized* labor (R).
4- New products render previous ones obsolete.
5- Creative destruction prevails.
6- Future research discourages current research.

Criticism of the Model

As mentioned before in the criticism of Romer's growth model, Aghion-Howitt's growth model is founded on the production of "physical goods". Therefore, it does not have a word to say about the developments about the output and employment in *services* sector, which has the largest share in GDPs across the world. A theory which does not explain the growth process in the services sector cannot be claimed to be a "general" theory of long-run growth. This failure alone is sufficient to reject Aghion-Howitt's model outright. In addition:

1- *"The model would gain richness and realism if capital were introduced, either physical or human capital embodying technical change, or R&D capital that effects the arrival of innovations."* (Aghion-Howitt;1992;349).
2- The obsession with concepts such as *equilibrium, unrealistic assumptions, and fictitious production relations* continues to be a serious barrier for the advocates of neoclassical ideology which prevents the followers of the ideology from escaping from their utopian world.
3- The assumption that "labor" is a "trade-able good" is not only an archaic assumption but also unethical and unrealistic. A laborer is a "human-being", not a tradable object or commodity. The sole purpose of all economic transactions is the satisfaction of this human-being.
4- There is no attempt to explain the source and functions of the *specialized* labor, which corresponds to "creative mental labor" in Chapter 2 of this book. Though a highly essential input of all technological progress, it swings in the air or stands in the loose sand. A specialized labor is a natural part of the human capital. Considering it as an "independent" and "different" input than human capital would only lead us to incorrect analysis and conclusions.

5- The neglect of capital-goods in the research sector contributes only to reduce the credit and the credentials of the model.
6- The assumption of employment of "unskilled-labor" only in the final sector producing consumer goods is just another oddity. If, according to the model, the development of new technologies is a continuous process, then the entire labor-force, including the laborers in the final goods sector should continuously strive to improve their skills in order to have efficient production. How can one expect the "unskilled" laborers to utilize the latest technologies efficiently?
7- Cost-saving innovations aimed at reducing the unit production cost of "existing" products are neglected. Yet, many innovations are introduced into the market are simply for this purpose. A growth model which does not explain cost-reducing innovations can only be a *"partial" growth model.*
8- The entire process for developing and employing the innovations between the varying time periods is far from explaining the real facts. Innovations are the natural result of competition, not a function of "rent expectations of next period".
9- The model cannot explain the differences in growth rates among countries.

The criticism of the sections extending the basic model of Aghion-Howitt is neglected on the simple grounds of saving time. That is because; when a model is constructed on loose arguments, like neoclassical models, there is no need to consider the rare qualities of its components scattered here and there; the construction is out of whack.

Grossman-Helpman: Foreign Trade and Growth

Grossman & Helpman focus on the endogenous technological change in their book titled "Innovation and Growth in the Global Economy" (1991) the subject of which can be described as the *"economic determinants of technological progress".* The book deals with; "... *the dynamic evolution of comparative advantage and the consequences of international trade in a world of global technological competition".* Technological innovations are "endogenous", because they emerge as a result of the intentional attempts of economic agents, responding to market conditions. These intentional behaviors are determined by the *"expected profit"* motive of the economic agents. Grossmann-Helpman; "... *concentrate on mechanisms that link the growth performance and trade performance of nations in the world economy."* (1991; Preface)

Chapters 1 and -2 outline the general information about their models. Solow's growth model is presented in Chapter-3. Chapters-3, -4 and -5, analyzes long-term growth in a single economy in isolation, i.e., in a closed economy. Starting with the Chapter-6, Grossman-Helpman focus on growth in an open economy. They examine the growth and foreign trade performance under the following headings:

1. A small and open economy (Chapter-6).
2. Intra-industry trade in high-technology (Chapter-7).
3. Regional spillovers (Chapter-8).
4. Global integration and growth (Chapter-9).
5. International transmission policies: Government intervention (Chapter-10).
6. Imitation: technology transfer (Chapter-11).
7. Product cycles: technology transfer (Chapter-12).

The main area of interest is the relationship between *"endogenous" technological innovations, foreign trade performance and growth.*

The Model

Suggesting that growth takes place not randomly, but as a result of many intentional attempts including various governmental policies, Grossman-Helpman try to find answers to two questions using the framework of *"general equilibrium"* analysis.

1. What are the reasons behind the long-term growth or the rise of per capita income?
2. What are the reasons behind the differences in growth rates?
   a) *"In the same period"* but *"in different countries"* and
   b) *"In the same country"* but *"in different periods"*?

Technology which is the source of growth is defined as *"a kind of knowledge"* and has the following characteristics:

1. *Technology is a non-rival good*: its use by someone does not preclude its others from using it.
2. *Technology is a partially excludable good*: its use by others can be partially precluded by means of patents.

General Assumptions:

1. Innovations are due to the inherent features of the economic system and related to profits.
2. Thanks to innovations found as a result of R&D, a monopolistic profit is obtained.
3. Savings are spent on R&D.
4. R&D has two fundamental purposes:

a) *To reduce the unit production cost* (process innovation).
b) *To produce a new product*:
   i) A completely new product (product variety) and
   ii) A product with a higher quality (quality upgrading).
5. *No capital-goods.*
6. *No human capital.*

Now we will briefly evaluate the subjects discussed in the Grossman-Helpman book one by one.

a-) New Products
i- Product Variety
The basic characteristics of product variety examined in the Chapter-3 of the book and are as follows:

1. The market is monopolistic.
2. Products in the market are imperfect substitutes.
3. The supply of the "new" products is infinite.
4. When the return on a new product falls to the discount rate level, investment ceases.
5. New knowledge (new technology) may be partially used by others (spillover effects).
6. In the final sector, technology has constant returns to scale (*CRTS*).
7. *Larger countries set aside higher levels of resources for the R&D. Therefore, both technological innovation/ and growth occur faster.*

ii- Higher Quality Products (Quality Upgrading)
Grossman-Helpman analyzes the supply of a given product with "higher quality" in Chapter-4 and determines that:

1. Given products are now of a "higher quality".
2. Quality of the given products may be increased infinitely.
3. Markets are imperfect.
4. The supply of a "given" product whose quality changes is not produced anymore.
5. New knowledge (new technology) may be partially used by others (spillover effects).
6. Competition exists.
7. Growth continues in a balanced manner.
8. *Larger countries set aside higher levels of resources for the R&D. Therefore, both technological innovation and growth occur faster.*

b)-Factor Accumulation

In Chapter-5, Grossman-Helpman focuses on the accumulation of factors such as capital goods and human capital. The particular characteristics of this chapter are:

1. Innovations are only produced by the labor.
2. Capital goods investment exists.
3. Capital goods may be homogeneous or heterogeneous.
4. Innovations increase the marginal productivity of capital goods.
5. Capital goods are used in the supply of either;
    1-) intermediate goods,
    2-) final products, or
    3-) both.
6. In the supply of intermediate goods, only labor (L) is used.
7. The final goods sector employs,
    1- labor;
    2- machinery, and
    3- intermediate goods.
8. Final products (Y) are consumed either by;
    1-) households or
    2-) producers.
9. *In the country with higher levels of qualified laborer, more innovations are introduced and the country grows faster.*

c-) A Small Economy Open to International Trade

Chapter-6 examines growth and innovation in a small economy open to international trade. The main assumptions specific to this chapter are:

1. There are two kinds of goods.
2. Goods are tradable.
3. International prices are "exogenous" and "given".
4. International interest rates are "exogenous" and "given".
5. Global demand is "perfectly elastic".
6. Supply is carried out according to the factor endowment in the country.
7. R&D activities in the country are at such levels that they will not have an impact on the global acquisition of human capital.
8. *Even a small economy may grow faster thanks to innovation and international trade, though it may not significantly influence global development.*

d-) Intra-industry Trade in the High-tech Products
The point of Chapter-7 is to demonstrate the dynamic comparative advantage in high-tech products. The model states:

1. There are two large countries and innovation take place in both of them.
2. Successes in R&D provide a comparative advantage.
3. There are two production factors. Their quantities are constant at the "stationary equilibrium" level.
4. There are two sectors with different levels of technological development.
5. The flow of ideas and knowledge is free between countries.
6. *International trade is shaped and the countries grow by gaining a comparative advantage gained through their R&D efforts and from the level of their technological development.*

e-) Geographical Knowledge Spill-overs
In Chapter-8, Grossman-Helpman analyzes growth and international trade with respect to the comparative advantage between two countries and by considering the historical dimension. The particular aspects of this model are:

1. There are two economies of different sizes.
2. The initial *R&D* structure of each country is different.
3. R&D results in one country contribute only to the overall knowledge enhancement in that country.
4. There is one production factor; labor.
5. Production inputs in both countries are the SAME.
6. Constant returns to scale (*CRTS*).
7. *At the initial stage, the country with the greater amount of knowledge acquisition gains the advantage. However, knowledge spillovers and the R&D results throughout history also affect the growth rate, not just the initial state.*

f-) Global Integration
In Chapter-9 global integration resulting from international trade and its contribution in growth process is discussed. According to Grossman-Helpman, four mechanisms have an impact on long-term growth.

1. International trade causes the transfer of technical information.
2. International competition pushes the firms to find "new" and "different" ideas and technologies.
3. International integration enables the markets to grow. Thus, international competition increases.

4. International trade between countries with different structures leads to the "re-distribution" of resources.

As a result of integration and increased competition, innovation increases and leads to long-run growth.

g-) Other Factors Effecting the Growth
Chapters-10, -11 and -12 analyzes the;

a) economic policies adopted by governments;
b) use of the technology by imitation; and
c) direct foreign investments or technology transfer through licensing;

in conjunction with innovations, spillovers and their contribution to growth and international trade and establish positive relationships between them.

Conclusions
In the long-run;

a- the amount;
b- the quality; and
c) the variety of the goods;

increase continuously. In other words, long-run economic growth is a continuous process. There are three factors enabling growth. These are:

1) Traditional factor accumulation;
2) Productivity increases as a result of technological innovation; and
3) Technology transfers to developing countries.

The model examined in the book attributes the knowledge spillover that results from "endogenous" technological innovations as the source of long-term growth. By virtue of technological innovation, contrary to the conventional expectation, long-term profit rates do not fall towards zero. Thus, Grossman-Helpman draws the conclusion that: *Productivity increase, based on "endogenous" technological innovations is the source of long-term growth.*

A Critical Evaluation
Once again, as mentioned before, Grossman-Helpman's growth model is founded on the production of "physical goods". Therefore, it does not have a word to say about the developments about the output and employment in *service* sector, which has the largest share in GDPs across the world. A theory which does not explain the growth process in the services sector cannot be claimed to

be a "general" theory of long-run growth. Grossman-Helpman ignore this fact in their model and this failure alone is sufficient to reject Aghion-Howitt's model outright.

In addition:

1. Grossman-Helpman establishes a relationship between growth, technological innovation and foreign trade from several different angles and contributes significantly to the science of economics. In addition to the endogenous technological progress, Grossman-Helpman introduces international trade into their growth model. This is, of course, a positive step forward. But, they make a serious error by assuming the production of a single, final good which reduces the credibility of the model.
2. Unfortunately, like other neoclassical economists, Grossman-Helpman is not able to free their mind from the "general equilibrium" obsession. This causes them to diverge significantly from the actual relationships encountered in the real world which they attempt to describe.
3. The assumption that prices are determined simultaneously in the product, factor and capital markets is a result of efforts to develop a "utopian general equilibrium" model.
4. The concept of "comparative advantage" dates back to D. Ricardo and it is essentially a misguided approach to international trade. Grossman-Helpman makes the same mistake in their analysis by building a model on comparative advantage. The mistake is that; *countries do not trade, but firms do.*
5. The adaption and the transfer of technologies by developing countries in the model are far from reflecting reality. The "spillover" effects of knowledge, thus of innovation applies, in general, to the *firms of developed countries*. Technology transfer to developing countries by means of direct foreign investment generally remains an unfulfilled expectation. And license agreements containing various restrictions seem to function as an obstacle rather than as a means to transfer technology to developing countries. Imperfections in the technology market and the increasing domination of global production for the benefit of firms in developed countries are serious barriers in increasing global competition and also seems to hinder the development of firms in the developing countries (see H. Gürak; 2013).
6. According to Grossman-Helpman, since monopolistic profits will decrease as the global competition increases, there may be a decline in the number of innovations. This is a mistaken idea. In fact, we should expect an increase in the number of innovations when global competition increases. That is because; the firms technologically falling behind (not the countries) will be driven out

of the global market, and thus excluded from global competition. As a result, the number of competitors in the global markets will decline over the course of time. Competition does not preclude companies from introducing new technologies, as Grossman-Helpman assume, on the contrary, it compels them to innovate.

G. Mankiw

In his book titled "The Growth of Nations" (1995), Mankiw asks some critical questions such as: What makes the rich countries richer than poor countries? How can the rich countries be sure to maintain their high standard of living? What can the poor countries do to catch up to the rich countries? (Mankiw;1995;275). He tries to find the answers by using the neoclassical model which he considers as "the natural place to start." "In evaluating the usefulness of the model in explaining growth", Mankiw finds the predictions of the model based on the assumption of/ a/ "*steady-state*" "broadly consistent with experience". These are:

1- "*In the long-run, the economy approaches a steady-state that is independent of initial conditions.*
2- *The steady-state level of income depends on the rates of saving and population growth.*
3- *The steady-state rate of growth of income per person depends only on the rate of technological progress.*
4- *In the steady-state, the capital stock grows at the same rate as income, so the capital-to-income ratio is constant.*
5- *In the steady-state, the marginal product of capital is constant, whereas the marginal product of labor grows at the rate of technological progress.*" (Mankiw;1995;277).

Under the title of "theoretical objections", Mankiw poses a very critical question: "*Is the neoclassical model a good theory of economic growth?*" (Mankiw;1995;280). One objection concerns, that in the steady-state, it is the exogenous technological progress which is considered the source of all growth. According to Mankiw, this growth model is not very illuminating in explaining growth, if the goal is to explain why standards of living are higher today than a century ago. Mankiw claims: "*... living standards rise over time largely because knowledge expands and production functions improve.*"

For Mankiw, attempting to explain the variations in growth rates in different countries and in different times is more challenging. In this regard, the objection to the assumption of the neoclassical model that "different countries use roughly the same production function at a given point of time" is not persuasive, says Mankiw. He claims: *"To say that different countries have the same production function is merely to say that if they had the same inputs, they would produce the same output."* (Mankiw: 1995; 281).

According to Mankiw, the more important question is: *"Can the model help to explain the wide variation in economic experiences observed throughout the world?"* (Mankiw: 1995; 281). He raises three reasons to challenge the validity of the neoclassical model: *"The issue at hand is not whether the neoclassical model is exactly true. The issue is whether the model can even come close to making sense of international experience."* (Mankiw: 1995; 282).

First of all, he says, the neoclassical model; *"... does not predict the large differences in income observed in the real world."* (Mankiw: 1995; 283). Secondly, regarding the convergence, Mankiw claims that the neoclassical model: *"... does not necessarily predict convergence. If countries are in different steady states, then rich countries remain rich, and poor countries remain poor. [...] if all countries have the same steady state and differ only in initial conditions, then the model does predict convergence."* (Mankiw: 1995; 284). Thirdly, on the rate of return, he says that according to the neoclassical model the rate of return in poor countries should be higher than in rich countries. Therefore, the poor countries should attract more capital. There is some evidence for rate of return differentials. *"The size of the predicted differentials, however, depends on the production function. The larger the elasticity of substitution between capital and labor, the smaller the return differentials."* (Mankiw;1995;289).

A New View of Capital

In the mainstream models or theories, the term "capital", normally, refers to the physical or tangible inputs of production such as machinery, plant, tools, etc. Mankiw considers "human skills" acquired through schooling or on-the-job training as "another form of capital", e.g., "human capital". He says: *"When applying the neoclassical model to understand international experience, it seems best to interpret the variable k as including all kinds of capital. Thus, the capital share, ∞, should include the return to both physical and human capital."* (Mankiw;1995;293).

It is an interesting point that Mankiw gives the impression that his model is not a neoclassical one. For example on page 295 he says: *"So far, my attention has centered on the neoclassical model."* Yet, he believes that by making some changes in the neoclassical model, he thinks the model can work. His production function is:

$Y = AK$

Accumulation equation:

$Ḱ = sY - δK$

Where s is the rate of saving, δ is the rate of depreciation of capital and Ḱ the increase in K. Together with the production function, this equation implies:

$Ŷ/Y = Ḱ/K = sA - δ$

As long as $sA > δ$, income grows forever. In other words, sA may lead to growth forever. Thus, cross-country differences in saving rates explain the differences in growth rates, says Mankiw.

> "The question is, how do we interpret the variable K in the production function, Y=AK? If K is seen as including only the economy's stock of plant and equipment, then it is natural to assume diminishing returns. [...] if we interpret K more broadly, then the assumption of constant returns to scale is more palatable" (Mankiw;1995;297).

Mankiw provides us the answer: *"The most appealing way of interpreting the endogenous growth model is to view knowledge as a type of capital."*

Mankiw distinguishes "knowledge" from the "human capital. *"Knowledge refers to society's understanding about how the world works."* So knowledge must be "general knowledge", not a specific type of knowledge to produce, i.e., "productive knowledge" as described in Chapter-1 and Chapter-2 of this book. *"Put crudely, knowledge is the quality of society's textbooks; human capital is the amount of time that has been spent reading them."* (Mankiw;1995;298).

Critical Evaluation

Once again, Mankiw's growth model is founded on the production of "physical goods". Therefore, it does not have a word to say about the developments about the output and employment in *service* sector, which has the largest share in GDPs across the world. A theory which does not explain the growth process in the services sector cannot be claimed to be a "general" theory of long-run growth. Mankiw overlooks this fact in his model and this failure alone is sufficient to reject Mankiw's model outright.

In addition:

1- The neoclassical models of growth or any neoclassical model is far from reflecting real transactions. The criticisms against the neoclassical parables stated in previous sections or elsewhere (see H.Gürak;2007) are valid here, too. As an addition, we shall evaluate the assumption of "steady-state. Mankiw believes that the neoclassical model is useful in explaining growth and the predictions of the model based on the assumption of *steady-state*[18], are claimed to be "broadly consistent with experience". If one is dedicated to a "faith" like the neoclassical ideology, using such fictitious and obsolete terms may not seem like a serious problem. As we know, neoclassical ideology has borrowed many terms, including the "steady-state", from cosmology in order to convince the readers that it is essentially a "positive science" like astronomy and physics. However, in time, as knowledge improved, these positive sciences have undergone radical changes. Most of the astronomers or physicists are keen on avoiding the use terms like "steady-state". But, the neoclassical economists seem to keep their scientific (!) positions and see no problem in using these obsolete terms.

2- According to Mankiw; *"... it seems best to interpret the variable k as including all kinds of capital. Thus, the capital share, ∞, should include the return to both physical and human capital."* Yet, he admits that human income known as the labor-income is not part of the capital income in national income accounts. Obviously Mankiw's assumption of a return on human capital does not reflect reality, at all. But, why does Mankiw and economists like him insist on referring to it as "human-capital" or a "return on human capital? Could it be ideological? Or do the neoclassical economists see no harm in breaking away from reality just to preserve their scientific (!) models? How much value and credit do such so called "scientific" (!) models deserve? Human capital is an inseparable part of the laborer. Unfortunately, once again, the reality is distorted to fit the "scientific" (!) model.

3- Mankiw makes his criticisms and present his ideas in such a way as if his model is not a neoclassical one. Yet, he attempts to present a "new" model by making some minimal changes to the neoclassical model, to make the neoclassical model work. Mankiw's model is a neoclassical one containing "minimal changes", not a truly "new" model of growth.

4- On page 300, Mankiw makes a remarkable confession in contrast to the role he assigned to knowledge. *"Even though knowledge is undeniably important for economic growth, theories of the creation of knowledge may be of little help in explaining international differences in growth rates."* A model which undermines

---

18 "Steady-state is a now-obsolete theory and model alternative to the Big Bang theory of the universe's origin. ... While the steady state model enjoyed some popularity in the first half of the 20th century, it is now rejected by the vast majority of professional cosmologists and other scientists," http://en.wikipedia.org/wiki/Steady_State_theory 2014-04-24.

the role of knowledge in growth in explaining cross-country growth rates is bound to be infertile. As shown in the previous sections of this book, the source of long-run growth in every aspect is the technological progress which is the product of creative mental labor.

5- Mankiw thinks: *"... the neoclassical model is still the most useful theory of growth we have."* (Mankiw;1995;308). In a sense Mankiw is correct; for so called "blackboard economics" with fictitious assumptions and relations that are far from reflecting the reality, the neoclassical theories and/or models are extremely useful to keep our brains working. But, unfortunately are of very limited practical value.

Middle-Income-Trap: End of Growth?

According to Eichengreen & et. al. (2013), extending their analysis given in an earlier paper, there is a corelation between growth rate and income level. According to the new data, they say, the middle-income-countries face slowdowns in growth at two modes; the first one being in the neighborhood of $10,000-$11,000 and the second one at $15,000-$16,000, (2005 purchasing power parity dollars), and are critical thresholds. After that threshold, a number of Middle-Income countries are at risk of a growth slowdown which is referred to as the 'Middle-Income Trap". However, they think that countries endowed with appropriate secondary and tertiary education where high-technology products account for a relatively large share of exports have the potential of avoiding the "Middle-Income Trap". That is because, after the threshold, the sustained long-run growth can only be realized through the employment of technological progress, which reduces the risk of a slowdown. The criterion for a growth slowdown is the sustained reduction in the growth rate for seven years.

> *"In analyzing the correlates of growth slowdowns, we found that slowdowns were positively associated with high growth in the earlier period (suggestive of mean reversion), with unfavorable demographics (high old-age dependency ratios in particular), with very high investment ratios (as if growth fueled by brute-force capital formation eventually becomes unsustainable), and with an undervalued exchange rate (as if countries with undervalued currencies have less incentive to move up the technological ladder out of unskilled-laborintensive, low-value-added sectors and thus find it more difficult to sustain rapid growth). These results were suggestive, and they were suggestive for China in particular."* (Eichengreen;2013;3).

Eichengreen & et. al. also claim to find some relationship between growth slowdowns and financial crises as well as changes in political regime. However, they are; "*... reluctant to push this evidence too far.. What is less intuitive is that* *"positive" regime changes – from autocracy to democracy – increase the likelihood of slowdowns."* (Eichengreen;2013;4).

Trade openness is another factor influencing the slowdown in growth. *"... external shocks might have a more powerful impact on the probability of a slowdown in more open economies."* (Eichengreen;2013;10).

In conclusion, Eichenreen & et. al. state:

> *"At some point, high growth in middle-income countries will come to an end. The low hanging fruit will have been picked, and high-return investments will have been completed. Underemployed labor will have been transferred from rural to urban sectors, while the demographic dividend will become a demographic drag. But this does not mean that a slowdown at a specific income level is inevitable. Not all countries are equally susceptible. Countries accumulating high quality human capital and moving into the production of higher tech exports stand a better chance of avoiding the middle income trap."* (Eichengreen;2013;14).

Critical Evaluation

Although; *"Not all countries are equally susceptible.",* Eichenreen & et. al. expect that: *"At some point, high growth in middle-income countries will come to an end."*, as claimed in the cited paragraph above. The arguments put by Eichenreen & et. al. have a good point regarding the critical roles of the qualification level of the labor-force and high technology exports. However, their arguments need some further qualification to account more comprehensively and satisfactorily for the explanation of underlying factors behind the so called "Middle-Income Trap".

It is true that after having reached an income level which is "lower than the average income level of the richest advanced countries", it would be difficult to raise it to a higher level, if the technology employed in the Middle-Income country is, predominantly, owned by "other" countries, especially the most advanced industrialized countries. The underlying reason behind this is partly due to the "imperfections" or "defects" in the global technological markets where the restrictive clauses of "patent agreements" prevent the efficient transfer of technology, and hinders increased global competition. The monopolist owner of the technology normally invests in the Middle-Income or "emerging" country to take advantage of the low-wage cost and/or other financial incentives/grants which reduce the unit-cost of production. And, it is, in principle, the labor-intensive parts of the output that are produced in the Middle-Income countries. If the wage rate rises above the "acceptable" limits and/or financial incentives loose their attraction, the technology owner can always move the production-site to another low-wage, financially high-incentive providing country, and there is nothing to stop this process. After all, we live in so called "liberal" countries.

If the Middle-Income country pursues a policy encouraging domestic firms to develop their own technology in order to become a global monopolist in technology market, the picture will be quite different for the Middle-Income country. Now the domestic firm can decide where to locate the production plant for labor-intensive parts, how to distribute the output supplied and where to accumulate the global income. The ownership of the technology (patents) makes a hell of a difference. Japan and S. Korea became high-income countries by encouraging the domestic development and ownership of technology. Therefore, they could climb the ladder of development rapidly. But countries like Turkey or Mexico which depend highly on foreign companies for the export of high technology products are, in a way, bound to be trapped in their present middle-income levels. Maybe some more or some less, but they can never attain the highest income levels unless they encourage the domestic firms to develop new globally competitive technologies, just like Japan and S. Korea did. Small countries like Singapore are not representative of all Middle-Income countries due to their specific circumstances and opportunities.

How high up the income ladder the Middle-Income countries climb depends mainly on the level of the quality of the labor-force or, as some say, the quality of the human capital. The higher the level of the qualifications of the labor-force, the higher the potential to earn a higher income would be. The imperative input is a high quality labor force with appropriate qualifications. However, one should note that "creating" a new technology is quite different to using the most advanced technology efficiently. That means, to climb the ladder of income the qualifications of the labor force are important, but more important is the "creative" abilities of the labor-force. And, last but not least, the structure of the "ownership" (control) of the technology is a critical imperative in order to reach the highest income levels.

In summary; four things are essential to climb the income ladder:

1- An appropriately "*qualified labor-force*" (human-capital) that can use the available technologies efficiently.
2- "*Creative*" minds to invent and develop sophisticated "*new*" products and/or methods of production.
3- The "*ownership*" and "*control*" of the technology.
4- The "*right*" and "*efficient*" institutional environment.

In Conclusion

Retrospectively, one observes that the economic growth analyses of the older "Classical" economists were quantitatively and qualitatively less sophisticated than are used today. But, nevertheless, their analyses appear to be more useful in grasping actual economic relationships when compared to the contemporary neoclassical analysis. For instance, the crucial factor of growth, "technological progress" which was "rediscovered" by the neoclassical economists in the 1950s, was already in existence in the analysis of the Classical economists as an "endogenous" factor. The same applies for the concept and role of "human capital", or rather the "qualifications of laborer", which was "also" rediscovered in the 1950s by the Neoclassical economists. However, more than 200 years ago, Adam Smith had written about the importance of the education of the labor force in regard to output and advocated that the children of poor families should have access to free education. At that time, there was no talk of a social state at all. Although technological progress was regarded as an "endogenous" process, being driven by internal forces, the Classical period economists had, unfortunately, failed to construct a coherent theory on long-run economic growth based on creative mental labor and technological progress.

After the 1870s, the ideological process of transforming what then was known as the "political economy" was initiated in order to pave the way for a *"purely scientific"* economic analysis, just like the one's found in the natural sciences e.g., astronomy with universally applicable laws. These new methods of analysis which were free of any human or subjective values gradually increased with ever growing sophistication of the mathematical models. But this caused many serious deviations from actual economic events and led economists to favor a "fictitious" approach in dealing with economic relationships. All historical developments, institutional settings, cultural-political relations and personal values had suddenly lost all their importance. Instead, the new emphasis was on how to engender "static equilibrium". Yet, "static equilibrium" has never actually taken place and, in all actuality, never will.

As a result, over a period of time the neoclassical ideology gained increasing attention and respect and acquired global recognition. Meanwhile, in 1938 the English economist Keynes succeeded in shaking the theoretical foundations of the neoclassical heritage in regard to such basic issues as full-employment and an automatic return to a state of equilibrium if everything was left solely to market forces. But the fundamental concept of long-run equilibrium remained unchallenged. In fact, Keynes was attempting to outline a path to long-run static equilibrium. Since the main focus was on acquiring equilibrium, the Keynesian theory had nothing at all to say about long-run growth. But then what? Does such a short term equilibrium theory, which has nothing to say about human capital and technological progress, deserve to be acknowledged as universal?

Actually, it seems to be more appropriate to regard the Keynes' theoretical analysis as a "different version" of the neoclassical short-run "static equilibrium theory". Keynes had never objected to the concept of static equilibrium and had nothing to say about long run developments. He developed no model correlating the qualifications of the laborer i.e. "human capital" and technological progress. His theoretical criticism of the neoclassical doctrine was not unjustified in regard to unemployment-equilibrium. But Keynes had "nothing new" to say about the political measures which intervene in the market to eliminate unemployment. At the time of the publication of his book in 1938, countries like Germany and USA were already implementing some "efficient" policy measures to eliminate the unemployment problem. In other words, the so called "Keynesian employment generating policies" which aimed to restore "full-employment equilibrium" was already being practiced in several parts of the world, before Keynes mentioned them in his book.

Some economists claim that the "Golden Age" in the application of Keynesian employment generating policies was in the post Second World War era until the time of the oil crises between 1945 and 1973. But this is not correct either, for that period was not characterized by any unemployment in developed countries, but, on the contrary, by a shortage of laborers, which led to the mass "imports" of laborers into Europe, Australia and New Zealand.

As to the prevailing neoclassical ideology; in spite of the fact that it basically preserves the "status quo" and does not appear to have diverged from its original path towards a state of equilibrium, it appears to have been subject to serious setbacks from the proponents of ideologies such as those of Solow et. al., with their "technological progress" and "qualified labor" approaches. Until the study of the causes of growth by Solow in the 1950s, the focus of economic analysis highlighted business cycles and the restoration of an "equilibrium", which, in fact, never existed. From the 1870s on until the 1950s, for about 80 years, the level of output was assumed to be determined by the level of employment of the two factors of production, labor and capital. Rediscovery of the effect of technological progress on growth of had caused the production function to change, followed by another rediscovery, that of "human capital". In other words, the original production function of the Neoclassical doctrine, $Y=f(K,L)$, first became $Y=f(K,L,A)$, and then $Y=f(K,L,A,H)$. According to Mankiw, the equation is: $Y=AK$.

L and H are in principle, the same thing, namely two sides of the same coin. Moreover, "A" originates from "creative mental labor" i.e., "L" and the knowledge created by mental labor is embodied in a physical form in the means of production. The neoclassical analyses inevitably led to many analytical misconceptions, misperceptions and miscalculations. Since the 1980s, many attempt have been made to indigenize the technological progresses and human capital in the so called "endogenous growth models". A lot of progress has been made, but there still seems to be scope and a need for further development in the theory of growth.

Concluding Remarks for the Chapter

When asked about the origin of *long-term* growth, "given" the natural resources, there is only one possible answer: "new investments" due to *new products introduced by technological progress which is produced by creative mental labor* (Gürak, 1993; 2000-b). Therefore, the "new theories" have to incorporate concepts like *mental labor* and *new products (technology)*, starting with value-price and growth theories and emphasizing their relationship to each other. Otherwise, economic theories will be "disabled" and "infertile".

Economic science has accumulated a good deal of knowledge and experience about the role and significance of these two key factors, particularly in economic theories related to growth. But it is hard to say that the expected realistic level has been reached yet. Studies on this issue should begin with a *value-price theory* before considering other theories, because the *value-price theory* is the basis of all economic theories.

Chapter-5
An Alternative Growth Model

Introduction

It is well known fact that the relationship between the labor and value-production was a top economic priority in the research conducted by the economists of Classical period. At that time, technological progress used to play an important role in their dynamic analysis and was treated as an "endogenous" factor. But, in spite of the important role assigned to it in their economic analysis, the Classical economists did not construct any satisfactory growth models which demonstrated the inter-relation between labor effort, technological progress and economic growth. Nevertheless, they were well aware of the significance and contribution of these to the long-run economic growth.

After the 1870s, long-run dynamic economic growth analysis began to be replaced by the "static equilibrium" analysis of the Marginalist and succeeding neoclassical doctrines. Inspired by the positive sciences like astronomy and physics, the primary aim of the new doctrines was to find new methods which could bring economies into "steady-state equilibrium" in order to be able to cope with the growing Marxist challenge. Leaving aside the attempts of Schumpeter in the 1930s and 1940s which re-emphasized the importance of technological progress in economic growth, the dominant economic growth models of this neoclassical heritage completely ignored this technological aspect, until the appearance of Solow's contributions in the 1950s. In other words, the economists of the neoclassical heritage according to the models they proposed had no idea at all of the impact of technological progress and how they affected the course of long-run economic growth.

The advocates of the neoclassical doctrine "rediscovered" technological progress and the fundamental role it played, thanks to the works of Solow. Using an analogy, they were very happy at finding the dog, which they neglectfully lost. After his "striking" rediscovery, Solow was awarded the Nobel Prize. However, Solow had some problems with the origin of technological progress and could not explain in his works how it emerged. But, he soon found an "ingenious solution" (!) to this problem by declaring that technological progress was an "exogenous" input to the system. They were being produced outside the economic system and "fell" into the economic system like "manna from heaven".

Despite all the shortcomings of the new theory, Solow's "rediscovery" engendered a renewed interest in the relationship between long-run economic growth and technological progress among economists. Later, economists also "re-discovered" the crucial role played by the *"qualities of the laborer"* (human capital) in this process. In the 1980s, new equilibrium models called *"endogenous economic growth models"* began to emerge, which considered technological progress as deliberate and conscious consequence of economic decisions.

All these neoclassical patch-up models also had many serious shortcomings in fully explaining the actual process of economic growth, just like all the other neoclassical models. These particular shortcomings have been discussed in the book entitled "Economic Growth and the Global Economy", by Gürak (2006). At present there does not appear to be any single economic growth model or a paradigm about which economists have displayed any consensus. There seems still to be a need and scope for further research and new ideas in the field of long-run economic growth analysis. The shortcomings of these prevailing theories are the driving force behind this present work.

The Purpose

The purpose of this work is to make a contribution to the theory of growth by presenting a "simple" growth model based on the concepts of "creative mental labor" and "technological development". The gifts of the nature, e.g., raw-materials and their market values are assumed to be a given. According to this hypothesis, the genesis of all the value added to the prevailing market values of raw-materials, thus the continuing source of all output growth and a nation's riches, is laborer, or rather, the level of qualification of the laborer. Therefore, it is the intention of this work to present an "alternative" long-run economic growth model based on a key concept namely the *"qualifications of the laborer"*. We encounter this concept, the "qualifications of the laborer, in every aspect and at every level when we undertake any long-run growth analysis. To put it more specifically, given the gifts of nature, *"creative mental labor"* of the laborer appears to be both the genesis of and the reason for the continuity of all long-run economic growth. To put it another way, technological advance, which is the indispensable ingredient of all long-run growth, is, in fact, the product of *"creative mental labor"*. The efficient use of these technologies is also an essential factor, but, its efficiency is also related to "mental labor". The qualities of this laborer can be divided into two general groups:

1. Technology-producing *"creative mind of the laborer"*;
2. Technology-using *"qualities of the laborer"*.

The technology-producing "creative mental labor" precedes the latter in importance in the formation of long-run economic growth.

With the application of technology, which is the result of mental labor, in the production process, raw-materials (gifts of nature) are transformed into "useful" products such as tools, intermediary inputs or consumption goods. To put it another way, the technology produced by the laborer is used to produce an output either for further production or final consumption. When new technologies are embodied in the tools/machinery of production or so-called capital-goods, they may either help to increase the "per unit time productivity" of employees, or introduce new products. Thus, the concept of the "marginal productivity" of capital (-goods) is nothing but a fallacy.

The share of the service sector in Gross Domestic Product (GDP) in today's economies is higher than the industrial or agricultural sector in terms of the value-added and employment. In spite of this undeniable fact, the growth models of the mainstream ideology seem to pay only lip service to the service sector contribution. The technological progress and the quality of the labor force affects the service sector's output differently than the industrial or agricultural sectors. The service sector analysis shall be presented later in a separate Chapter under the title "Growth in the Service Sector".

The probability of the acceptance of an "alternative" growth analysis as presented in this section is not high, especially by the proponents of the neoclassical doctrine who rigidly adhere to their equilibrium analysis based on fictitious assumptions and models. We know very well by experience how conservative minded, in fact ignorant; the neoclassical proponents are to "other" ideas.

By the way, some economists may think: Why should we bother with an alternative theory of some unknown Turkish economist while there are plenty of works by well-known Anglo-Saxon economists? If these economists decide to leave aside their "prejudices" and try to evaluate this "alternative" growth theory with, say 10 percent, of the degree of the objectivity and tolerance they gave to the Anglo-Saxon or Western growth models they may actually encounter some *"logical, consistent and more realistic explanations"* for long-run economic growth other than those found in the neoclassical models. If we have respect for "science" and "scientific methods", we have to be more open-minded and less-prejudiced.

"Productive" Factors & "Production" Factors or Inputs

There were only two factors of production considered by the mainstream economic theories; viz. labor (L) and capital (K). Due to the developments in economic theories since the 1950s, two factors are now added to the original two; namely technological change, (A), and human capital (H). In this section of this work, the reader will be presented with a different approach to these issues which differs from those prevalent at this time. This approach claims that there are only "*two productive factors*" of production, but "*many inputs or factors of production*".

Productive Factors

There are only two "productive" factors capable of producing use- and/or exchange-value as discussed in the Chapters 1 and 2.

1- Laborer (L) (physical as well as mental).
2- Nature (the entire ecological system)

Factors (Inputs) of Production

In this section all required inputs of production are the factors or inputs of production. For instance, along with labor and capital goods, all raw-materials, the energy used, buildings, tools, in short, every item necessary for a required output are factors (inputs) of production. In contrast to the orthodox equilibrium theories, capital goods are not assumed to be productive; on the contrary, they are used to increase the productivity of the laborer employed in production. The common factors (inputs) of production are:

- Laborer.
- Raw-materials.
- Intermediaries (semi-finished goods).
- Energy inputs, water, etc.
- Capital goods (machinery, tools).
- Consultancy services.
- Post-production marketing and sale efforts.
- Transport-insurance.
- Management.
- And all other inputs required for the particular output.

Factors (inputs) of production can be subdivided into two broad groups:

1- Laborer (L).
2- Other inputs (X_i).

"Productive" Factors and Value-creation

There are only two "productive" factors; nature and labor. Nature is productive in the sense that it is capable of supplying products with "use-value" without any external intervention. These products range from directly consumable products such as vegetables and fruits to the basic inputs of production which in turn are transformed by labor. The productivity of the nature is closely associated with environmental conditions. Nature, in modern societies generally, does not supply products, which are directly consumable. In order to be consumable contemporary products have to be "*transformed*" into "useful" products by some form of labor.

It is only after being processed by labor these supplies from nature are transformed into products with an "exchange-value".

The labor-time spent could range, from a "simple" labor-time, say transporting the apples from a garden to the market place, to a more "complex" labor-time requiring higher qualifications in transforming nature's products into semi-finished or finished products. For instance, the raw form of a chair is the tree, and it is transformed by labor into a useful product with an "exchange-value". It is a primary law of physics that nothing in nature disappears completely and nothing is created without using some form of available input; natural supplies only change forms through labor. That is to say, nature supplies the basic inputs of all output and laborer converts them into the other forms demanded by the consumer. Assuming that the inputs of nature are a given, the source of all "exchange-value" is generated by the physical and mental inputs of laborers.

Following this line of reasoning, an attempt will be made below to construct a simple growth model based on labor and laborer. It will not be the aim of this simple model to give an exact account of actual complex economic relations. But, rather, it could be used as a precursor to pave the way for more realistic models in the future. The main purpose of the simple model is to show that the original source of all created exchange-value, technological innovation and long-run growth is laborer, or, more specifically, *"creative mental labor"* of the laborer. Therefore, the reader is asked to bear in mind the mental labor aspect of the model all times.

The Genesis of Growth

In the subsequent section, the concept of labor will be categorized under two headings, in order to get a clearer insight into its contribution to the growth process:

1. *Physical labor.*
2. *Well-qualified labor or human capital.*

The qualification of the labor or synonymously the labor-quality or the human capital can further be divided into two categories:

2-a: Technology-using labor.
2-b: "Technology-creating" labor or "creative labor".

1- *Physical labor* (L^b) refers to all kinds of basic physical activities such as walking, drinking, holding, etc. Such activities are similar to those made by other living beings in order to survive in their environment. Such activities can be initiated by basic instincts. But, nevertheless, the control center of every type of activity is the brain, and in the absence of these mental directives, living beings could not survive. Even the most basic activities are initiated by instructions from brain. Therefore, there is always some degree of mental contribution involved in every stage of existence. The concept of physical labor, in our case, simply means the carrying out of these sets of instructions, which result in the basic co-ordination of the human physical system.

2- *Laborer's level of qualification* (L^n): As mentioned before, nowhere, there is a labor-force without some degree of qualification (i.e. without any form of education, training, skills) i.e. a laborer endowed with only of purely physical labor. Every laborer possesses some degree of qualification that is referred to as human capital. Therefore, it would be a serious error to label the labor force in terms of the two artificial categories presently being used i.e. as being "qualified" and "unqualified". This error will not be repeated in this work, unless in appropriate and exceptional cases.

There are five basic factors determining the level of qualifications of labor and its productivity:

1- The individual's "natural" capabilities and talents.
2- The general level of the knowledge in society.
3- Formal and informal education.
4- Learning-by-doing.
5- Experience.

Any increase in any one of the factors mentioned above would in turn raise the qualifications of the labor force. Every capable individual enjoys a certain level of ability. In other words, every individual possesses a certain degree of the five factors stated above. At the least, in a modern society some capabilities would be acquired through informal education from the family and others from environmental factors. In advanced societies learning would be more visible as almost every individual attends the school for at least 12 years. Therefore, the claim that every individual possesses some degree of qualification (human capital) is not easy to challenge. The qualification level of the laborers in developing countries, on the other hand, seems to be much lower due to inadequate education/training facilities.

There is a direct relationship between the level of a laborer's qualifications and the standard of education in a particular country. The higher the level of qualification of the labor force, the higher the expected individual or total level of wealth would be. On the other hand, regardless of the level of the individual's natural talents, if the nation's general development level is below the global contemporary level, productivity per unit time employed would be expected to be lower than the global standard. That is to say, the level of the quality of the labor force and the productivity of a nation is directly related to the general level of the accumulated knowledge and the quality of the labor force. As mentioned before, creative mental labor is the genesis of all the added value and long run growth, but the present level of productivity (the efficient application of technologies in production), depends on the quality level of the labor force.

Some of the productivity growth may be attributed to practice at work. In his third model, Lucas (1988) claims that the necessary capabilities for production are acquired through the "learning-by-doing" process. The quality of work done increases as the hours spent at work increases, given the same technology. Workers are so specialized at practicing the job that over time the cost of production is expected to fall while and output to increase. Thus, per unit time productivity would increase with increasing practice at work.

Experience is a concept that has a wider meaning than simply "learning-by-doing". The impact of experience, the importance of which seems to be underestimated, is also significant in the development of mental capability and thus to value generation. For instance, a doctor or a nurse may learn surgical procedure step by step by learning-by-doing with a given technology; experience would make them more efficient in terms of the general issues related to simple surgical procedures. Similarly, the contributions of an experienced person in making a decision on a strategically important issue can be extremely vital. More experienced individuals are more likely to make wiser and less faulty decisions, which may have a great impact on their present and future work. In a similar fashion, a more experienced teacher or security officer who has learned the job through the "learning-by-doing" method is likely to be more productive at work after several years of experience.

2-a: Technology using labor of varying quality levels: Those who produce new technologies and those who employ these technologies are, in general, not the same laborers. There is always a need for qualified labor to efficiently employ the existing technologies. In other words, the degree of utility of a technology depends on the qualities of existing labor force. For instance, if the labor force is not properly equipped with the knowledge necessary for efficient production, it would be impossible to produce, say, airplanes or automobiles, or at least not of the same quality. Therefore, the qualities of the labor force are important for the efficient production and the total wealth of society.

2-b: "*Creative*" or "*productive*" *labor* is the source of all the added value accrued from any creative activities or changes, which involves those activities beyond the basic sets of instructions sent from the brain governing purely physiological activity. In modern societies, creative labor is, in general, employed in R&D departments in search of new ideas. Research funds are normally employed to finance the creation and development of either" new products" or "new production methods" which in turn are used to produce the available goods or services at a lower cost.

Naturally, it is not only the highly educated, but also those with a relatively lower level of formal education can contribute to the creation of new ideas. One way or another, the new ideas or the knowledge required to raise the productivity per unit time employed and the long run growth is always a result of the creative abilities of human beings. Accordingly, all technological developments necessary to increase both individual and total productivity stem from the creative mind of individuals.

Some Basic Assumptions

1- The determining factor of long run growth is technological progress (A), which is a product of creative mental labor, (L^y)
2- Nature, one of the two "productive factors", supplies the necessary inputs for production, while laborer transforms them into useful products.
3- L^n denotes the average laborer endowed with average level of qualification (human capital). L^n, in its turn, determines the level of the labor force's productivity. L^n includes both, the creative mental labor L^y, and qualifications of laborers, L^k, using the technology in production.
4- Let us denote the qualifications of average laborer in Turkey by L^n_{TR}, and in Germany by L^n_D. The present situation due to the differences in qualification level can be shown as $L^n_D > L^n_{TR}$.
5- There are two countries or producers (TR,D) with an equal quantity of laborforce. Further assume that the quantity of "physical labor-time" spent in one day is equal in both countries $(L^b_{TR}=L^b_D)$. Under these circumstances, the wage rate in both cases should be equal $(w_{TR}=w_D)$.

However, the situation described above will change as soon as we take into account the differences in qualifications of the two labor forces. Assume that the average qualifications of labor force in D are better than in TR $(L^n_D > L^n_{TR})$.

Naturally, the labor force in D would be more productive than the labor force in TR and enjoy a higher wage rate ($w_D > w_{TR}$). The difference between the qualifications of the labor forces is due to a relatively higher degree of accumulated knowledge in D, better educated and trained labor force, and the technological as well as the institutional development level.

An Attempt at an Alternative Growth Model

Firstly, before attempting to construct an alternative growth model to the mainstream models, we shall consider a simple stationary economy with two producers, two consumers and two goods to exchange. "Only physical labor" is employed and no growth takes place.

Then, we shall consider a simple growth model with the same two producers and two goods in which one of the laborers develops a "tool" by utilizing her "creative mental abilities which increases her daily output. Increased output or in similar fashion increased productivity implies growth in the quantity supplied.

Based on the results of the two simple models, the growth analysis will continue with the study of growth in a "real economy" where the qualified labor will be included in the model to show its impact on growth.

Initial Case: A Simple "Stationary" Output, Exchange & Distribution Model

The purpose is to analyze the barter-exchange relationships, along with the individual and the total consumption level, using the "physical labor-time employed" approach. The growth does not occur since the economy is stationary. The structure of the model facilitates the study of income distribution together with growth.

Assumptions:

- There are only two producers and two consumers, Leyla, (L) and Maria, (M).
- Only two products are produced and consumed; X_1 and X_2.
- Consumer preferences are the same.
- No accumulation. All output is consumed on the same day of production.
- No money. Barter-exchange takes place.
- Only "physical labor" is used in production, L^b.
- Since "creative" mental labor, L^y, has not yet been introduced, there is no new technology (A) developed nor any "means" of production (capital-goods) K,

has been produced. Thus, there is no need for any qualified labor, L^k, for the efficient employment of any technology.

The production function is:

$$Q = f(L^b_L, L^b_M)$$

L^b_L, denotes Leyla's, and L^b_M Maria's physical labor. Initially, both Leyla and Maria enjoy "the same level of qualification" that of simple "physical labor" ($L^b_L = L^b_M$). Each of them work 10 hours a day and produce two different products (X_1 and X_2). Leyla's daily output is 4 units of X_1 that of Maria is 2 units of X_2, and both have identical tastes and preferences. At the end of each day, they exchange products worth 5 hours of labor-time (2 X_1 = 1 X_2). The outcome is:

Leyla's output	4 X_1	10 hours / day
Maria's output	2 X_2	10 hours / day
Total output	$QT = qL + qM = 4 X_1 + 2 X_2 =$ 20 hours / day	
Leyla's consumption	$CL = 2 X_1 + 1 X_2$	
Maria's consumption	$CM = 2 X_1 + 1 X_2$	
Total consumption	$C_tL,M = 4 X_1 + 2 X_2$	

Both, Leyla and Maria, spend equal quantities of labor-time and, as a result, consume equal quantities. The exchange is "fair" in terms of labor-time employed and both enjoy an equal quantity of use.

In the absence of any "mental" labor contribution, which means there are no technological innovations, so no growth would take place, because the production capacity and tastes are as stated. The existing system is capable of only maintaining the status quo of any existing production and exchange relations. Equilibrium exists but there is no growth.

For growth to take place, both the output and consumption have to increase. For output to increase there is need for "creative mental labor" that means, new technologies have to be introduced. There has to be either innovation introducing *a new production method for a "given" product*, or *entirely new products with new production methods*. In the following models, we shall assume that, given the product, *a new production method* increases output.

A Simple "Short-run" Growth Model: 1

Innovation & Growth: A Given Product but a New Production Method

Additional assumptions:

1- By utilizing her "creative mental" abilities, Leyla develops a tool (a capital-good) which increases the output of her labor-time employed. Leyla's labor input now is no longer purely L^b, but L^n. Her productivity per unit time employed is increased.

2- Supply creates its own demand, i.e. supply and demand is in balance. Every additional item produced is consumed, but the exchange-relationships will have to change, cet. par.

Leyla increases her daily output from $4\ X_1$ to $8\ X_1$ with the employment of the "innovation" developed by her creative mental ability. Initially, Leyla's labor had no qualification level; that is $L^n_{L,t} = 0$.

But now;

$$L^n_{L,t+1} > L^n_{L,t}$$

And;

$$q^L_{t+1} > q^L_t$$

For Maria, the initial conditions are still valid.

$$L^b_{M,t+1} = L^b_{M,t}$$

And;

$$q^M_{t+1} = q^M_t$$

The new total production function is:

$$Q = f(L^n_L; L^b_M)$$

The new technology (A) developed by Leyla's "creative mental labor" (Ly) is embodied in a material form in the means of production and help to increase her productivity.

Since preferences and working-hours have not changed and there is no third party with whom to enter into an exchange-relation, so in order for the entire output to be consumed, the production and exchange relationships have to change:

Leyla's output	$8\ X_1$	10 hours/day
Maria's output	$2\ X_2$	10 hours/day
Total output	$Q_{t+1} = q^L_{t+1} + q^M_{t+1} = 8\ X_1 + 2\ X_2$	= 20 hours/day

The "fair" exchange ratios according to the "labor-time spent" approach would be as follow:

Leyla's consumption	$C^L_{t+1} = 4\ X_1 + 1\ X_2$
Maria's consumption	$C^M_{t+1} = 4\ X_1 + 1\ X_2$

Total consumption $\qquad C^{L,M}_{t+1} = 8 X_1 + 2 X_2$

Both producers continue to work 10 hours a day, as in the initial case. But, due to the "new technology" developed by Leyla's creative mental labor, total output is increased (growth occurs):

$$Q_{t+1} > Q_t$$

As is personal consumption:

$$C^L_{t+1} > C^L_t \qquad \text{and} \qquad C^M_{t+1} > C^M_t$$

Since, by assumption, there is no supply-demand imbalance, every output supplied is consumed, but exchange-ratios have to vary. Although the increase in total consumption is entirely due to Leyla's contribution, the other consumer, Maria, who has so far contributed nothing, now benefits from the new situation just as much as Leyla. To put it differently, Maria, who makes no contribution to total output growth, benefits just as much as Leyla from the new situation. Leyla's creative productivity is, in a sense, being penalized, while the "stationary" position of Maria is rewarded. This kind of exchange relationship calculated in accordance with "labor-time employed" criterion does not seem "fair", at all.

Denoting growth with the symbol "g", the growth depends on the qualification level of Leyla's labor, Ln, or to be more precise, on her "creative" mental labor, Ly.

$$g = f(L^n)$$

Assuming Access to a "New" Market

In the model presented above, exchange relationship was based on the assumption that the exchange-partners consisted of only two individuals. Now, we assume that additional items supplied by Leyla are sold to a new buyer in another market, while the exchange relationship between Leyla and Maria remains the same.

Assume that Leyla exchanges the additional 4 units of X_1 with a third person and receives 3 units of X_3 in return. Under the new circumstances, the total output of the community, consisting of Leyla and Maria will increase, along with Leyla's output and consumption, while the position of Maria, as in the previous case, remains unchanged.

Leyla's output $\qquad 8 X_1$
Maria's output $\qquad 2 X_2$
Total output $\qquad Q_{t+1} = q^L_{t+1} + q^M_{t+1} = 8 X_1 + 2 X_2 = 20$ hours/day

After trade with a third individual:

Leyla's consumption	$C^L_{t+1} = 2X_1 + 1X_2 + 3X_3$
Maria's consumption	$C^M_{t+1} = 2X_1 + 1X_2$
Total consumption	$C^{L,M}_{t+1} = 4X_1 + 2X_2 + 3X_3$

In other words;

$$Q_{t+1} > Q_t$$
$$C^L_{t+1} > C^L_t$$

But;

$$C^M_{t+1} = C^M_t$$

Under the new circumstances, Leyla's consumption along with the total consumption of the community increases (grows) as a result of Leyla's contribution. As for Maria, who made no contribution at all to the growth process, consumption remains unchanged.

The "egalitarian" exchange relationship in terms of labor-time employed principle is not valid anymore. The new exchange relationship seems to be more rational and more realistic promoting further innovations which would facilitate further economic growth. This is, actually, what really happens in actual economies.

A Simple "Long-run" Growth Model: 2

Technological Productivity Growth: New Product and a New Production Method

In the simple model presented above, we studied how growth occurred and how it affected exchange relationships, with a "given" product. Now, we shall assume that Leyla, by utilizing her "creative" mental labor, develops an *"entirely new product"* highly likely produced by a "new production method", denoted as X_4, in addition to the previous increase of the product X_1. Let us first analyze the supply and exchange relationship between Leyla and Maria.

There are two producers and two consumers, along with as in the previous models, no lack of "effective demand".

The production function is:

$$Q = f(L^n_L; L^b_M)$$

10 units of the new product (X_4) are produced, and the entire output is consumed in a domestic market consisting of Leyla and Maria. As before, Maria will

be the major beneficiary of growth as a result of egalitarian exchange based on the labor-time employed, although she has made no contribution at all.

Leyla's output $\quad\quad$ $4 X_1 \quad + 10 X_4$
Maria's output $\quad\quad$ $2 X_2$
Total output $\quad\quad$ $Q^T_{t+1} = 4 X_1 + 10 X_4 + 2 X_2$

The outcome of the "fair" barter-exchange based on the "labor-time employed" approach would be as follows:

Leyla's consumption $\quad$ $C^L_{t+1} = 2 X_1 + 5 X_4 \quad + 1 X_2$
Maria's consumption $\quad$ $C^M_{t+1} = 2 X_1 + 5 X_4 \quad + 1 X_2$
Total consumption $\quad\quad$ $C^{L,M}_{t+1} = 4 X_1 + 10 X_4 + 2 X_2$

The output as well as the consumption is higher. That means that growth has occurred.

"New Markets" for the "New Product"

Assuming a state of affairs where half of the new product is exchanged in another market for a product demanded by Leyla. Thus, Leyla's consumption level will be improved, while Maria's will remain the same.

Let's the following explanatory example: Assume that Leyla's "new" product, X_4, is used in trade with a third party and 5 units of X_4 is exchanged in return of 6 units of X_5. The new production and consumption relationships will be as follows:

Leyla's output $\quad\quad$ $4 X_1 \quad + 10 X_4$
Maria's output $\quad\quad$ $2 X_2$
Total output $\quad\quad$ $Q_{t+1} = 4 X_1 + 10 X_4 + 2 X_2$

The outcome of barter-exchange:

Leyla's consumption $\quad$ $C^L_{t+1} = 2 X_1 + 1 X_2 + 5 X_4 + 6 X_5$
Maria's consumption $\quad$ $C^M_{t+1} = 2 X_1 + 1 X_2$
Total consumption $\quad\quad$ $C^{L,M}_{t+1} = 4 X_1 + 2 X_2 + 5 X_4 + 6 X_5$

To summarize;

$$Q_{t+1} > Q_t$$
$$C^{L,M}_{t+1} > C^{L,M}_t$$
$$C^L_{t+1} > C^L_t$$

But;

$$C^M_{t+1} = C^M_t$$

Now the output as well as the consumption is higher implying that growth has occurred.

Conlusions to be Drawn from the Simple Growth Models

The simple growth models studied above clearly show that the cause of all productivity increase is (technological) innovations which are the product of creative mental labor. Along with the increase in *"productive knowledge"* or *"knowledge on production"* i.e. new technology, not only economy grows but also individual and total wealth increases. Since there seems to be no upper limit to the creativity of the human mental abilities, there seems to be no barriers, for now, for the long run growth of any economy.

Negative developments affecting environmental issues may reduce or exhaust completely the quantity of the necessary inputs of production, which, in turn, would bring an end to the growth process. But, it would not be irrational or illogical to expect that the necessary precautions would be introduced in time to prevent such a disaster. The decline in the global reserves of oil and coal should not emerge as a serious problem in the energy sector, because creative mental labor is capable of producing alternative energy sources. For instance, borax and hydrogen may be sources for future energy.

To summarize, one can easily claim that, assuming the supplies of nature are a "given" then the source of all exchange-value creation that was accumulated in the past, is being accumulated at present and may be accumulated in the future is the result of creative mental labor which constantly develops innovations.

Limit of Short-run Growth

Assume that a firm produces goods (X), and all resources, human, physical and financial, are employed at an optimum level, that is maximum technical efficiency prevails and profits are maximized while costs are minimized. What are the options for firms if they desire to increase their profits?

1- *Horizontal expansion:* One of the options is to find *new markets* for their products. In order to meet the potential demand for "given" goods from the new markets, the producer may have to make "expansive" investments abroad" or increase the exports. As long as demand grows, the income and thus total profits of a firm will continue to increase, cet. par. However, given the product, there is always an upper limit for demand. As the market for a given product begins to saturate, the strength of demand will begin to decline and eventually halt. After that point, the output could only aim to meet a demand caused by depreciation and population growth, given the existing purchasing-power. The impact of population growth on output growth can only be marginal. Thus, the growth process without "new products" is sooner or later bound to come to an end.

2- *Wage cut:* Another option for a firm to increase its profits is to reduce the wages paid to employees, cet. par. Thus, the share of wages in the added value accrued will drop and the share of profits increases, while the total added value remains unchanged. This option may prove beneficial for the wage-reducing firm, but if all the firms introduced wage cuts at the same time it would likely produce opposite results for the economy. It would reduce the total demand and the total added value produced, implying a negative growth. Therefore, what is beneficial for one firm is not necessarily beneficial for the other firms or the economy.

3- *Reducing the cost of imported inputs:* Given the technology and an optimum efficiency level in production, the only possibility to reduce costs seems to be paying less for the imported inputs of production. But, the suppliers of these inputs would naturally be reluctant to sell their inputs for a lower price, which would lead to their profits being reduced, cet. par.

There is always a limit to growth with "given" products because as the markets approach saturation point, the strength of demand is bound to decline and eventually stop. Population increase may support the growth process to some extent, but not sufficiently, in the long run. For the long run growth, the introduction of "*new products*" is imperative.

Technological Innovations and Producers

We produce and consume a wide range of products, some of which like shelter, clothing and food are called the 'needs' and some like automobiles, TV-sets and washing-machines are called the 'wants' or 'desires' to make life more comfortable. All these items of consumption are produced by profit driven firms aiming to fulfill our needs or wants; the latter may also be referred to as 'derived needs'. The items of consumption are constantly increasing which is due to 'new' technologies. In other words, 'new' technologies or 'technological progress' is continuously influences our way of living and habits.

Why are the competitive producers in continuous search for new products or new methods of production? What is the importance of 'technological progresses in competitive markets?

Three major reasons can be outlined to explain why the profit-maximizing firms in competition seek for technological productivity growth introducing "new products" or production methods".

1. To be ahead in competition.
2. To maximize long-run profits.
3. Not to lag behind competitors.

The main goal of every commercial firm in competitive environment is first to survive competition and then maximize the profit rate and long-run profits in long-run. In sort-run, the commercial firm may increase profits by reducing some production costs through some non-technological measures. But, in the long-run, the only way to keep profits from falling to zero is to introduce new products and production methods that is technological productivity growth.

a) *Competitive advantage:* By reducing the unit costs of a "given" product by introducing a new production method, the firm would gain cost advantage against competitors. That could imply either higher profit rate if the price remains unchanged, or price advantage if the cost reduction is reflected in reduced price. If entirely new products are introduced by new technology, the firm is expected to enjoy higher than average profit rate and access to new markets.

b) *Monopoly advantage:* The owner of the new knowledge to produce *new product/production process* normally applies for a patent, which facilitates to enjoy monopolistic advantages until the others catch-up. Meanwhile the *expected* profit rate of technology owner is, quite likely, above the average market profit rate. In the absence of such expectations, the producer may not have sufficient incentives to finance the risky R&D process to introduce new products.

c) *Defensive strategy:* Assume that the competitor(s) of a firm acquire cost advantage or aim to gain a larger share of the market by introducing new technology. If the firm in question still employing the old technology does not take a counter measure to reduce production cost, it would risk losing the market and eventually cease to exist. In order to survive the competition, it has to catch-up with others either by inventing a competitive technology or by transferring it through a patent/license agreement. The cost of ignorance is withering away from the market for good.

The Importance of Innovations

What should be the characteristic features of a contemporary firm to survive and to maintain a competitive edge in global markets in the long-run? In view of the above statements, one can draw the following conclusions:

i. Have access to a labor-force endowed with required qualifications to *create* and to *use efficiently* the new technologies, along with appropriate technological infrastructure.

ii. If the required technology is transferred, not created by internal resources, then the firm should have the appropriate capabilities to *adapt* and *further develop* the new technology.
iii. Employees of the firm should get continuous intra-firm professional training to keep pace with the most recent developments. Decision-makers of the firm should be open to all kind opinions and promote participation in decision-making.
iv. Optimum technical or economic efficiency should be one of the major priorities of the firm.
v. Sort- and long-run expectations and goals of the firm should be both, rational and realistic with regard to global facts and developments.
vi. The firm must have appropriate dynamics to take the necessary risks and steps at right time and place.

Technological Innovations and Consumers
From the point of consumers, technological productivity growth, in general, provides to major utilities:

1. Cost-reducing technological innovations of "given" products, may reduce the sale price of products, which would lead to increased consumption, cet. par.
2. "Entirely new" or "improved quality" products are introduced at the service of consumers.

In both cases, the consumers benefit from technological productivity growth.

Employment and Technological Productivity Growth
Technological innovations usually display two different impacts on employment:

1. Causing job-losses, thus reducing employment.
2. Creating new jobs, thus contributing to employment.

In retrospect, one can frequently encounter cases where people meet new technologies with suspicion, or even display hostile resistance to change. The underlying reason of this behavior is the fear to lose jobs, in other words, loses income. However, again in retrospect, one can observe that while, on one side, causing job-losses, the new technologies provide, on the other side, new employment opportunities in many "new fields". Which is more beneficial for the mankind; keeping the available jobs with "given" technologies or promoting new technologies and new employment opportunities? Historical developments have already answered this question: the "new" wins over the "old".

Conclusions

According to the findings of this section, laborer appears as the only production factor capable of adding value, assuming the supplies from nature (the basic inputs of production) as both stable and a "given". Labor transforms nature's supplies into useful products to be sold in the markets. All "means-of-production" are transformed natural inputs aimed at increasing labor-productivity. Thus, labor, or rather the qualification level of the labor is the source of all the added value of a product as well as all new technology required for any growth to take place.

As to the qualification level of the labor force; it is proportional to the technological development level of the country in which they live and also to the quality of the education acquired. The increase in the number of technological innovations and in the level of general welfare of a nation is only achieved by the contribution of the creative mental abilities of the labor force.

In any theory in regard to value-creation and growth, it is essential to distinguish between the qualified and non-qualified laborers with regard to their contributions to production. Laborers endowed with qualifications (or similarly with human capital) may be "*technology-using*" laborers, which is important for the efficient use of any given technology. If the qualification level of laborer falls short of expectations, the outcome would inevitably lead to inefficiency and a waste of resources. Therefore, it is not only necessary to be endowed with a certain level of qualification but it is also imperative to have the "right" endowment in order to avoid inefficiency in the supply of products. Alternatively, given the technology, there is a direct relationship between the capabilities of the laborer and the level of advancement of a country.

Above reference was made to "*technology-using*" laborers. In the long-run where technological progress takes place, it is imperative to have access to "*technology-creating*" laborers in order to increase the total and individual wealth in an economy. As we know, technological progress is the source of all long-run growth while *creative mental labor* is the source of all technological progress.

A Note on Labor, Innovation & Value-Price Theory

Without hesitation we may claim that nowadays it is a commonly accepted by economists that technological progress is an indispensable input of long-run growth. However, the related and critical question is: what is the role of technological progress in the formation of the market price? How reliable and scientific are the mainstream value-price theories which do not even pay lip-service to the innovations created by "creative mental labor-power"?

As we know, the value-price theory is the foundation stone of all other economic theories. The entire economic system functions in response to the signals coming from prices. The wage-rate, the profit-rate, new investments, consumption demand, exports, imports many other issues are all influenced by the price level. We can only have more realistic theories with explanatory and predictive value if we have access to a more realistic value-price theory based on the innovations created by "creative mental labor". (see; H.Gürak, Economics, 2012).

Chapter-6
Short-run Growth in the Real Economy

In the simple growth models presented in Chapter-5, we studied growth and distribution relationships under barter-exchange conditions. Barter-exchange relations do not reflect the actual relationships adequately but help us to comprehend the basic relationships in a simple way. The analysis is believed to serve its purpose.

In this Chapter, an attempt will be made to reflect the growth process with the assistance of extended-models in a more realistic way in comparison to the mainstream models. Some generalized and abstract statements, as well as some fictitious assumptions and economic relationships will certainly crop up; but the intention is to make as realistic an analysis as possible in order to reflect real economic relationships.

The growth process was studied under two sub-categories, as described in Chapter-2.

> *Short term Growth:* (with a "given" technology).
>> 1-a) *EE* and/or *TE* improvement.
>> 1-b) Production for new markets.
>
> *Long-run Growth:* ("New" technologies).
>> 2-a) A Given Product, but a New Production Method.
>> 2-b) A New Product and a New Production Method.

We start with the short-run growth analysis with a "given" technology in this section.

1- Short-run Growth: A "Given" Technology

1-a) EE and/or TE growth

The crucial assumption in the short term is production with a "given" technology, which implies an absence of technological innovation. Under such circumstances, if there is a lack in economic efficiency (*EE*) or technical efficiency (*TE*), in other words, if *EE* and/or *TE* are not at an optimum level, the output can be increased until it reaches the optimum level. But, once the optimum level is reached, growth ends, and the economy is in equilibrium.

Table: 6-1 The causes of short-run growth

Efficiency Growth 1- Economic efficiency (EE) 2- Technical efficiency (TE)	Short-term	A **"Given"** technology

At the optimum capacity utilization and efficiency level where there is no unemployment, output should be somewhere on the $Y_1 Y_2$ curve, as indicated in Figure: 6-1. At point Z, either EE or TE or both are not at maximum capacity. Output can be increased until it reaches the Y_1-Y_2 curve. But, then growth comes to an end.

Figure: 6-1 Short term growth with a "given" technology

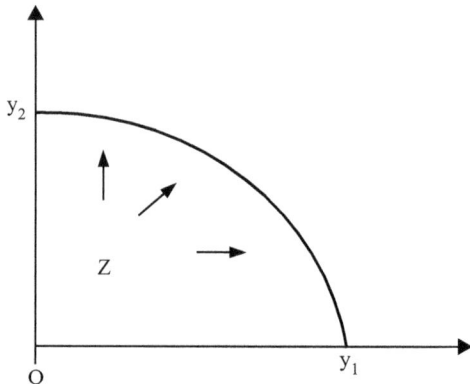

Assumptions:

1- Technology is a given (T).
2- Non-labor inputs of production (X).
3- Laborers required for production (L).

The production function is:

$$Q = f(L; X_i)$$

X_i denotes all inputs of production such as energy, raw-materials, marketing, the means of production, etc., excluding labor.

1-*Economic efficiency* (EE) increase: Given the technology, EE covers all attempts to reduce production costs and increase income aiming to maximize profits. EE can be stated as below.

$$EE = TR : TC = (P_s Q_s) : (wL + p_i X_i)$$

p^i denotes input prices, p^s output price, Q^s the quantity supplied, X^i the input quantities, w wage level and L number of employees. EE growth implies a positive change in its value.

$$EE \text{ growth} = \Delta TR > \Delta TC \quad \text{or} \quad \Delta VA > \Delta TC$$

For instance, acquiring inputs at a lower cost or a rise in the sale price of a product would imply *EE* growth.

If $TC \downarrow$ cet. par. $VA \uparrow$, $\pi \uparrow$ and $EE \uparrow$
Or, $p^s \uparrow$ and $TR \uparrow$ cet. par. $VA \uparrow$, $\pi \uparrow$ and $EE \uparrow$
Or, $\Delta TR \uparrow > \Delta TC \uparrow$ cet. par. $VA \uparrow$, $\pi \uparrow$ and $EE \uparrow$

2- *Technical efficiency* (*TE*) increase: *TE* is a concept related to the more efficient exploitation of physical inputs, given the technology. In other words, it refers to the efficient utilization of physical inputs and plant capacity. For instance, capacity under-utilization would imply technical inefficiency.

$$TE = \text{Quantity Supplied / Supply Capacity} \quad (\text{"given" the technology})$$

When *TE* ratio is equal to one (*TE=1*), the quantity supplied is at the maximum attainable level with the given technology. When the *TE* ratio is less than one (*TE<1*), it implies that the physical resources can be utilized more efficiently and the physical quantity supplied can be increased. As *TE* moves toward one, physical productivity will continue to improve.

The rate of profit which is the driving force of production for an enterprise will be maximized when *EE* and *TE* reach their highest possible levels. In other words, for maximum profitability level *EE* and *TE* have to be at their maximum values.

Short term growth (g^s) is primarily a function of *EE* and/or *TE* in the pursuit of profit maximization.

$$g^s = f(EE, TE)$$

Thus, the short-run growth with a given technology is the consequence of improvements in EE or TE.

1-b) Extended Production for New Markets

Additional assumptions:

1- Domestic EE and TE are at the optimum level.
2- There is demand for the "given" product from abroad.
3- No laborer constraint.

In order to meet the foreign demand, new production capacity (new investment, I), is required. Since *EE* and *TE* are at optimum levels, then short-run growth with a "given" technology would be a function of *new investments*.

$$g^s = f(I)$$

New investment is a function of the expected rate of profit (r^E).

$$I = f(r^e)$$

However, after a while, the market will saturate and growth will stop while the profit rate eventually drops to zero.

Eight Methods to Increase Short-run Growth

It is possible to increase productivity even in the absence of any technological innovation or progress. For instance, an enterprise may increase productivity by restructuring the production method, thus making more efficient use of the given physical resources. In a similar fashion, productivity with the given technology can be increased by improving the skills of the labor force or by increasing the capacity utilization level or improving health and security measures, etc. Table: 6-2 shows some of the impacts of efficiency growth on the value added (VA), rate of profit (r) and the share of profit in VA, while causing the share of the wage (w) in VA to fall, cet. par.

Table: 6-2 Relationships between growth, added value (VA) and profitability (r)

New technology	Cause of efficiency (micro-productivity) growth	VA/K	VA/L	r	π/VA	w/VA
No	Restructuring production	↑	↑	↑	↑	↓
No	Increasing capacity utilization	↑	↑	↑	↑	↓
No	Multiple shift-work	↑	↑	↑	↑	↓
No	Reallocating the resources	↑	↑	↑	↑	↓
No	Improved general education	↑	↑	↑	↑	↓
No	Learning on-the-job & experience	↑	↑	↑	↑	↓
No	Improved safety & sanitary	↑	↑	↑	↑	↓
No	Democracy at plant-site	↑	↑	↑	↑	↓

Certainly one can add other methods to achieve the desired increase, but we will suffice here with eight methods, to set some examples. In the meantime, we assume that the average enterprise in the market is in a globally competitive state and the socio-economic as well as the institutional infrastructure is at the global standards.

1-Re-organizing the Production

One of the well-known methods of increasing the efficiency at the plant site will be by re-organizing the production, or as A. Smith once pointed out, by re-organizing the division of labor. In the well-known example, A. Smith, suggested that if a pin factory had adopted a "division of labor", it might produce tens of thousands of pins a day whereas a pin factory in which each worker attempted to produce pins, from start to finish, would produce very few pins. The reason would be the concentration of each worker on only one part of the production, instead of each worker performing all the tasks associated with pin production.

Another way to increase the efficiency can be achieved by re-organizing the non-laborer inputs in such a way which would reduce any waste. For example, a ready-wear producing enterprise may, after the decision to re-organize, start to produce each part of final product in one floor, instead of producing the parts in different floors, which, highly likely, might increase the efficiency or productivity. Or in an office of an accountant, re-organization of the circulation of documents may improve efficiency.

An enterprise selling spare parts for cars and keeping an inventory may increase the efficiency by working with zero stock of inventory and thereby increase the efficiency. Examples like these may be easily multiplied.

Let us give an example with hypothetical figures, now. Assume that:

$w_t = 10$ TL   (wage level)
$L_t = 100$ (workers)
$p_t = 4$ TL   (price)
$FC_t = 1000$ TL   (fixed-costs)
$OC_t = 2$ TL (incl. wage)   (operating costs)
$q_t = 800$ pieces   (output)
$TC_t = FC + OC * q = 1000 + 2 * 800$ $= 2,600$ TL (total cost)
$TR_t = p_t * q_t$ $= 4 * 800$ $= 3,200$ TL (total income)
$\pi_t = TR_t - TC_t = 3200 - 2600$ $= 600$ TL (size of profit)
$r_t = \pi_1 / TC_1 = 600 / 2600$ $= \sim \% 23$ (rate of profit)
$VA_t / L_t = (wL_t + \pi_t) / L_t = (1000 + 600) / 100 = 16$ TL/per L

VA_t denotes the added value at the time "t". Assume that the re-organization leads to an efficiency increase resulting in additional output of 200 pieces. Total output is now 1,000 pieces.

$\Delta q = 200$
$q_{t+1} = 1,000$
$\Delta TR = p * \Delta q = 4 * 200 = 800\ TL$
$TR_{t+1} = p * q_{t+1} = 4 * 1,000 = 4,000\ TL$
$TC_{t+1} = 1000 + (2 * 1000) = 3,000\ TL$

This efficiency or productivity increase, without any additional cost of production, would increase the size and rate of profit as well as the added value per worker.

$\pi_{t+1} = TR_{t+1} - TC_{t+1} = 4000 - 3000 = 1,000\ TL$
$r_{t+1} = \pi_{t+1} / TC_{t+1} = 1000 / 3000 = \sim \% 33$
$VA_{t+1}/L_{t+1} = (wL_{t+1} + \pi_{t+1})/L_{t+1} = 1000+1000/100 = 20\ TL/per\ L$

2- Increased Plant Capacity Utilization

Economics textbooks of mainstream ideology assume, in principle, that there is no underutilized plant capacity that is the output is realized at optimum levels of capacity. Unfortunately, this assumption is far from reflecting the reality in enterprises. Plant underutilization is not an exception but a rather common fact of life. Especially in periods of economic crises, the underutilization of plants reaches to extreme levels. Plant underutilization which indicates to the underutilization of available resources is a serious economic problem both in developed and developing countries.

Let us give an example with hypothetical figures, now. Assume that:

$w_t = 5\ TL$
$L_t = 100$
$p_t = 4\ TL$
$OC_t = 2\ TL$
$FC_t = 1,000\ TL$
$q_t = 600$
$TC_t = FC_t + (OC_t * q_t) = 1,000 + 2 * 600 = 2,200\ TL$
$TR_t = p_t * q_t = 4 * 600 = 2,400\ TL$
$\pi_t = TR_t - TC_t = 2400 - 2200 = 200\ TL$
$r_t = \pi_t /TC_t = 200 / 2200 = \sim \% 9$
$q_t / L_t = 600 / 100 = 6\ pieces/ per\ L$
$VA_t / L_t = (wL_t + \pi_t)/ L_t = (500 + 200) / 100 = 7\ TL / per\ L$

Assume that the plant reaches its full plant utilization level of 700 pieces.

$q_{t+1} = 700$
$\Delta q = 100$

How would be the effect on total income, rate of profit and total profit?

$TC_{t+1} = 1{,}000 + (2 * 700) = 2{,}400 \text{ TL}$
$TR_{t+1} = 4 * 700 = 2{,}800 \text{ TL}$
$\pi_{t+1} = 2800 - 2400 = 400 \text{ TL}$
$r_{t+1} = 400 / 2400 = \sim \% \ 17$
$q_{t+1}/L_{t+1} = 700 / 100 = 7 \text{ pieces / per L}$
$VA_{t+1}/L_{t+1} = (wL_{t+1} + \pi_{t+1})/L_{t+1} = (500 + 400) / 100 = 9 \text{ TL/per L}$

As a result of full plant capacity utilization, the rate of profit (r) increases from nine percent to 17 percent, while output per worker (q/L) increases from six to seven and per worker added value (VA/L) grows from seven TL to nine TL.

3 – Shift Work

Another method of increasing the efficiency with "given technology" is to increase the intensity of use of capital-goods by introducing shift-work. This would enable the investor to increase the total output, thus reduce per unit production cost with the "given amount" of capital-goods, cet. par. In other words, due to shift work, per unit fixed cost (FC/q), including capital-goods and the plant cost, would decrease which, in its turn, would result in increasing the rate of profit (r) and the total added-value (VA). In order to achieve these results, there should be no shortages of neither required workers or necessary inputs of production or any other kind of obstacle of production. By assumption, the wage level and the price level are constant and demand no shortage of demand for the additional output.

According to the hypothetical figures given below, although the output per laborer ($q_t/L_t = q_{t+1}/L_{t+1}$) remains the same, the added value (VA), the rate of profit (r) and total output ($Q = q_t + q_{t+1}$) increases.

Assume that initially the enterprise has one shift work only.

$p_t = 15 \text{ TL}$
$OC_t = 2 \text{ TL}$
$FC_t = 30{,}000 \text{ TL}$
$w_t = 20 \text{ TL}$
$L_t = 1{,}000$
$q_t = 5{,}000 \text{ pieces}$

$q_t / L_t = 5$ pieces per L
$TC_t = FC_t + (OC_t*q_t) + (w_t*L_t) = 30,000 + (2*5,000) + (20*1,000) = 60,000$ TL
$TR_t = p_t * q_t = 15 * 5,000 = 75,000$ TL
$\pi_t = TR_t - TC_t = 75000 - 60000 = 15,000$ TL
$r_t = \pi_t / TC_t = 15000 / 60000 = \% 25$
$VA_t / L_t = (wL_t + \pi_t) / L_t = 20*1000 + 15000/1000 = 35$ TL per L

Now further assume that the enterprise introduces two shift works. In this circumstance, the enterprise has to employ additional 1,000 workers. As a result of shift work, output increases by the additional amount of 5,000 units ($\Delta q = 5,000$ pcs.) How would the total income, rate of profit, total profit and output per worker be affected?

$TC_{t+1} = FC_t + [(OC_t*q_t) + (w_t*L_t)]*2 = 30,000 + [(2*5,000) + (1,000 * 20)] *2$
$= 30,000 + [(10,000 + 20,000)] *2 = 90,000$ TL
$TR_{t+1} = 15 * 10,000 = 150,000$ TL
$\Pi_{t+1} = 150000 - 90000 = 60,000$ TL
$r_{t+1} = \pi_{t+1} / TC_{t+1} = 60000 / 90000 = \sim \% 67$
$q_{t+1} / L_{t+1} = 10000 / 2000 = 5$ pcs / L
$VA_t + 1/L_{t+1} = (2*wL_t + \pi_{t+1}) / L_{t+1} = (2*20000 + 60000) / 2000 = 50 TL/L$

As we see, VA per L increases from 35 TL to 50 TL. Since the real wage level is constant, cet. par. the owner of the enterprise will be the beneficiary of this development and the share of profit (π) in the added value (VA) will increase. The reason of this is the reduced fixed investment cost per unit of output.

4- Restructuring (Relocating) the Resources

Another method of increasing the productivity without any technological progress involved is to re-allocate the available investment resources to areas with higher added value per worker or per unit of investment. For example, the real wage level in the developed countries is relatively higher than the real wage level in developing countries. Therefore, develop country enterprises prefer to invest in technologically advanced sectors where they can obtain higher profits and added value per worker in comparison to what they can obtain in so called labor-intensive sectors such as textiles or ready-wear. This is a completely rational behavior from the point of view of the capital owner.

Since the real wage level in less developed countries like Turkey will have to rise in accordance with the economic development, they will inevitably become less competitive in cheap-labor-intensive sectors and eventually have to leave the field to countries with lower real wage level. Therefore, countries like Turkey have to make rational adjustments in right time as the economy grows.

Let us give an example with hypothetical figures, now. Assume that:

$w_1 = 1$ TL
$L_1 = 1,000$
$P_1 = 4$ TL
$q_1 = 1,000$ pcs
$FC_1 = 1,000$ TL
$OC_1 = 2$ TL

TR denotes the total income, TC the total cost, π the size of profit and r the rate of profit.

$TR_1 = p_1 * q_1 = 4 * 1000 = 4,000$ TL
$TC_1 = FC_1 + OC_1 * q_1 = 1,000 + 2 * 1,000 = 3,000$ TL
$\pi_1 = TR_1 - TC_1 = 4000 - 3000 = 1,000$ TL
$r_1 = \pi_1 / TC_1 = 1000 / 3000 = \sim \% 33$
$VA_1 / L_1 = (wL_1 + \pi_1) / L_1 = (1000 + 1000) / 1000 = 2$ TL/L

Further assume that the enterprise decides to leave the sector and make a new investment by the same amount of capital as before which is expected to bring in higher added value VA. Let us see what happens to TR, TC, π and r, cet. par.

$P_2 = 5$ TL $\qquad P_2 > P_1$
$q_2 = 1,000$ pcs $\qquad q_2 = q_1$
$FC_2 = 1,500$ TL
$OC_2 = 1.5$ TL
$TC_2 = FC + OC * q = 3,000$ TL or $\qquad K_2 = K_1$
$TR_2 = p_2 * q_2 = 5 * 1000 = 5,000$ TL
$\pi_2 = TR_2 - TC_2 = 5000 - 3000 = 2,000$ TL
$r_2 = \pi_2 / TC_2 = 2000 / 3000 = \sim \% 67$
$VA_2 / L_2 = (wL_2 + \pi_2) / L_2 = (1,000 + 2,000) / 1000 = 3$ TL/L

Although the enterprise invests exactly the same amount of capital ($K_1 = K_2$) total profit (π_2) and rate of profit (r_2) is higher. Since by assumption there is no change in the real wage level, the share of the profit in added value (VA) changes in favor of the capital owner.

5- Improving the General Education/Training Level

Observations show clearly that there is a close correlation between the "general" education and training aimed at improving the knowledge level of the individuals and the "general" productivity growth. The main reason of this is that the higher the "general" education-training level of the individuals, the easier it is for them to further develop their skills and to learn new things which would further increase their productivity in general. Since developed country decision-makers are fully aware of this correlation between education-training and productivity growth, they continuously and consciously support all kind of related investment in human beings. While the individuals in developed countries acquire at least twelve years of schooling, the situation in many developing country is quite frustrating. A significant share of the population still lack even the primary education and what a large share of the individuals get in terms of education-training is still far from the contemporary standards of technological development level.

6- *Learning-by-doing and Experience*

In the previous sub-chapter above, we claimed that the better educated individuals have the capacity or the potential to learn faster the new" things required in the production process, among other things. The best way to make use of this potential is to learn the necessary complementary knowledge in the supply of products at plant site by what is called learning-by-doing process which may also be referred to as the "intra-firm training". This is because, regardless of the standards of "formal" education, nobody can perform at full personal capacity in production right after the schooling period. There is often a huge gap between what is thought as useful knowledge at the formal educational facility and what is practiced during the production process. Therefore, learning-by-doing directly affecting the productivity level is of vital importance for a competitive producer. And most important of all is the "mentality" of the firm as well as the laborers that learning is a life-time continuous process. Not only individuals but also enterprises have to learn continuously in order to survive long-run competition.

Experience, a concept usually not encountered in the Western economic doctrines is directly related with the time of employment. Experience can neither be taught at formal educational facilities nor can be gained as such in time. Experience is a combination of the laborer's specific skills, education and working years. It is a state of mentality to improve personal productivity in the field, and often requires decision-making on critical issues. A car mechanic or a salesman or an engineer or an airplane pilot with five years of experience at work and an absorptive mental capacity would be expected to be more productive compared to others with less working years and/or less mental capacity.

7- *Improving the Health and Security Facilities*

Improving the health and security measures of the employees at the production unit may contribute significantly to productivity increase, without making any technological changes in the production process. If a production unit lacks the necessary precautionary health and security measurements, or if they are not at sufficient standards, then there would be a potential risk to lose some working hours or days of the employees, due to physical or mental illness. In other words, insufficient or inefficient health and security measures would highly likely result with undesired economic losses, not to mention the personal compensation costs.

In the case of absence of an employee due to poor health or security conditions the producer may decide to employ temporary worker, but this would imply higher labor costs per unit of output and less profit for the employer. The more the working hours or days lost, the higher the loss of the employer would be cet. par.

If the physical units of production say capital goods or the building suffers damage, this would imply once again higher unit production cost due to the decline in total output and reparation cost. And this would imply deteriorated productivity and loss of profits.

8- Increased Participation at Decision-making

What is suggested here has to do with participation in decision making on "how" the production process is carried out, rather than participating at management level decisions. To be more specific, the workers directly involved in the related part of production should also have a say on performing the specific duty. If realized, this approach is expected to increase the productivity as well as the responsibility of the workers since they will be directly responsible on the quality and quantity of the product supplied. By allowing the workers to offer ideas on reorganizing or restructuring the specific processes under their responsibility, they might be able to avoid monotony and frustration performing the same duties all the time which would affect the productivity negatively. In addition, participation in decision making at specific levels directly concerning the workers is highly likely to lead to the introduction of new ideas increasing the productivity. Minimizing the waste of resources is also highly likely to follow. Various studies around the world indicate that active participation of workers has many positive impacts ranging from improved quality, increased productivity and minimized waste of resources all of which contribute to higher amount of added value.

Chapter-7
Long-run Groth in the Real Economy

Technological (Macro) Productivity Growth

In the short-run growth models in Chapter-6, the technology employed (T) in production and the qualification level of the laborer (Ln) was assumed to be a "given". From now on, we shall focus on a case where the "creative" labor (Ly) enters the picture and paves the way for long-run growth by introducing "new" technology. In other words, the long-run growth takes place due to the technological innovations and the improvements in the qualifications of the laborers.

In Chapter-2 of this book, the concept of "technological productivity growth" or simply "long-run growth" refers to growth due to innovations. Now, we have a new dimension; "new technology" or "technological progress". In contrast to a "given" product and a "given" method of production, the "new technologies enable us to consume a wider range of "new" products or buy the "given" products at a lower price due to cost-reducing "new" methods of production.

The long run growth will be studied under two headings, see Table: 7-1, as in Chapter-2, but this time the analysis will be more detailed. We start first with refreshing our minds shortly on the technological productivity growth concept.

Table: 7-1 Long-run growth

"Given" product-new production method "New" product & "new" production method	Short-run Long-run	"New" technology

Assumptions:

1- "Creative" mental labor *(Ly)* introduces new technologies *(A)*.
2- Fair competition.
3- *EE* and *TE* at an optimum level.
4- No imbalance in supply-demand *(S=D)*.
5- No shortage of qualified laborers.

a) A "Given" Product, but a "New" Production Method

"Given" the product, the most rational and effective behavior of firms to improve competitive strength and profits is to introduce a "new" unit cost reducing production method. In a competitive environment, the producers would make every effort to produce the unit at less cost, in order to gain a cost advantage against their competitors. The producers who ignore this are bound to disappear from the market.

Assume that one of the competitors succeeds in reducing unit costs by employing a new production method. The reduction in unit cost may be due to one of the factors mentioned below:

1- "Given" inputs, but output increases. TC veri, $VA \uparrow$
2- Output increases faster than the increase in inputs. $\Delta VA \uparrow > \Delta TC \uparrow$
3- Output increases while inputs decrease. $\Delta VA \uparrow > TC \downarrow$
4- Inputs decrease while output remains the same. VA veri, $\Delta TC \downarrow$

The producer who achieves a cost advantage by using new technology would have three options.

1- To reduce the sale-price.
2- To increase the profits by maintaining the same price level.
3- To follow a price-policy combining both the options mentioned above.

However, there is always an upper limit to growth with a "given" product though the producer may enjoy a competitive advantage due to the new technology, which reduces unit costs. Sooner or later, the markets are bound to saturate, as in the short-run analysis, the profit rate fall and growth rate will diminish and eventually stop.

The production function is:

$$Q = f(L^y, L^k, L^b, X_i, A)$$

Or simply:

$$Q = f(L, X_i, A)$$

L, includes the labor-force with a sufficient level of qualification (L^n) to employ the new technology (A) efficiently. As we know, technological innovations are introduced by L^y in order to increase the productivity of the labor force and are embodied in the physical products. Therefore, there is only one "productive" factor in the production function; L.

Technological innovations (A), are internalized (embodied) in the physical products employed in the production process (X_i) and are a function of "creative" mental labor (L^y), the technological development level (T) and R&D investments (I^{R-D}):

$$A = f(L^y, T, I^{R\text{-}D})$$

Assuming that there is no depreciation, the growth function would be:

$$g = f(I^A)$$
$$I^A = f(r^e)$$

I^A denotes "new" investment due to technological innovation, L^n the laborers required to use the new technology, and r^e the expected profit rate. The laborer requires further qualification for the efficient employment of the new technologies. Therefore, for the growth process to be complete, the qualification level of the laborer has to be increased in addition to a new level of investment.

As displayed in Figure: 7-1, demand would increase somewhat, assuming that there is a price fall as a result of the cost-reducing technological innovation. As the product is assumed to be given, demand is expected to decline with the saturation of markets, which would cause the profit rate to decline, which, in its term, would cause a decline in the investment rate. And the growth process will eventually come to an end, when the markets saturate.

Figure: 7-1 Limits to growth and profit rate (cost-reducing technology)

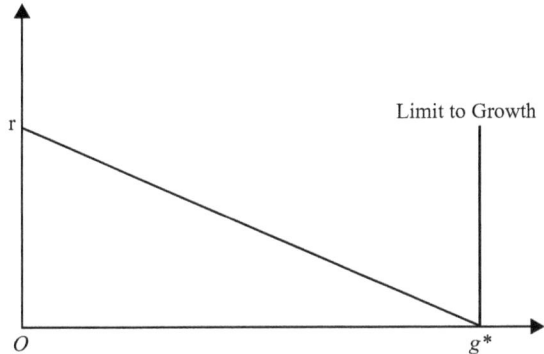

b-) New Products

The distinguished feature of this kind long run growth is the introduction of "new products" usually accompanied by "new production methods". Long-run and sustainable economic growth can only be realized by these kinds of "innovations". The long-run growth rate is normally subject to fluctuations, even to crises; but neither the growth process ceases, nor the profit rate tends to fall to zero in the long-run. Often, the ecological problems become the major subject of economic debates. But, nevertheless, the long run growth process continues.

In contrast to approximately 3,000 hours worked annually 50-60 years ago, an employee works about 1,500-2,000 hours nowadays and produce much better quality and more variety products. The only reason for these developments is continuous *innovations*, i.e., the supply of new technologies, or, to be more specific, a continuous supply of *new products*.

New products require new investments (I), which naturally generate a new demand. As a result, the initial profit rate never tends to fall to zero. If the profit rate had a tendency to fall below the market average, the firms would not have the incentive to undertake new investments. As a result of new technologies, not only the profit rate tends to be positive, but also the consumption options for consumers increase.[19]

Long-run growth is a dynamic process without a foreseeable limit, though it may, from time to time, experience some fluctuations. There seems to be no upper limit to the creative capabilities of mental labor in creating new products and/or production methods; nor does there seem to be an upper limit in the "lust" (demand) for new products. Under such circumstances, there would be no grounds to expect markets to saturate, nor for the growth process to end in the foreseeable future. The profit-rate would follow a trajectory with peaks and troughs, in accordance with technological innovation.

Now, we shall study how the introduction of new products influence the growth process, assuming EE and TE are at an optimum level. As before, $S=D$ by assumption.

Production function:

$$Q = f(L, X_i, A)$$

Assuming no depreciation of the means of production, the long run growth function would be:

$$g^L = f(I^A)$$
$$I^A = f(A, L, r^e)$$

As displayed in Figure: 7-2, the profit-rate would never fall to zero level given the constant supply of technological innovations. However, it is expected to

---

[19] A new product can be an entirely new one or an existing product with new features. For instance, although the service supplied by mobile phones is the same as traditional phones, the former is considered as a "new product", due to some qualitative improvements. Accordingly, "smart" TV sets are also considered as "new products".

fluctuate. To put it another way, in the initial phase of a new technology, the owner will enjoy "monopoly" privileges in the market and is very likely to obtain a higher profit-rate above the average for the market. In time, the competitors are assumed to introduce similar technologies and products, which would cause the monopoly profits to decline. But, a wave of new technologies would again facilitate monopoly profits, and the process would continue in this manner. Therefore, the long run growth rate is never expected to decline to a zero level permanently.

Figure: 7-2 Probable fluctuations in the rate of profit.

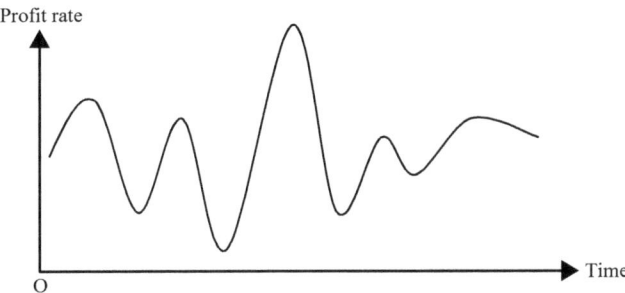

The reason for the long-run profit rate not to fall to zero and the source of continuous long run growth is "innovations". Invention and application of new technologies is an irreversible and uninterruptible process of economic life in progressive societies. Incessant innovations do not only serve the interests of producers but also of laborers. As a result of innovations, laborers work less hours, produce much more and better quality products. In addition, the real income per capita increases, facilitating more consumption and improved living standards. In spite of some setbacks and objections, people in general, welcome innovations.

To summarize; as long as demand for new products does not cease, new products will continue to be developed by creative mental labor, thereby increasing output and consumption, thus paving the way for a continuous growth process in the long-run.

Growth: Reconsidered - Both in the Short- & Long-run

In the models above, we first studied the short term growth process in the absence of technological innovations, and then long-run growth with technological innovations. Naturally, in real economies the individuals do not make decisions in line with the models. Producers normally begin a production process in regard to their long-run expectations, though they often face some unexpected incidents in the short term and might have to make some critical decisions. In fact, the long-run outcome consists of short term processes. The question is: Is it possible to a have a production and growth function covering both periods, i.e., the short and long-run.

Regardless of the time aspect, whether short or long-run, the inputs of production, or, in line with the orthodox terminology, "the production factors", are always "the same".

$$Q = f(L, X_i)$$

However, the case for the growth function is somewhat different. Because, we now have to take into account the growth process in both the short and long-run.

$$g^L = f(EE, TE, I^A)$$

When EE and TE are optimum, then:

$$g^L = f(I^A)$$
$$I^A = f(L, r^e)$$

I^A, denotes both, new investments due to innovation while L denotes the laborers required to efficiently realize production, and r^e denotes the expected profit rate.

The Concrete Effects of Innovations

In this section, we shall study the effects of innovation on growth in terms of the increase in the added-value along with on income-distribution and the profit-rate.

By assumption, there is no shortage of demand for the output, i.e. supply creates its own demand.

Given Product – New Method of Production

Let's see how a new process technology, i.e. a new method of production, influences the added value (VA) e.g. growth and the profitability, given the product. Assume that the firm produces car tires while employing laborers (L) and using X_i unit inputs. The production function is;

$$Q = f(L, X_i) \qquad i = 1,2,\ldots, n$$

Further assume that this producer introduces a new method of production and starts using less of the input X_5 though the quantity of output is the same.

Assuming that all prices of inputs, prices of output and the wages remain unchanged, given the demand, total cost of production (TC) will be lower now due to new technology;

$$Q_{t+1} = Q_t$$

But,

$$TC_{t+1} < TC_t$$

t denotes the time. Though the total quantity supplied (Q) and the capital required (K) did not change, the added-value (VA), total profit (π) and the rate of profit (r) increased, due to the new technology requiring less of the input X_5 now. By the way, total capital (K) includes all kinds of expenditure on production, including the laborer (L) and all inputs of production as well as the capital-goods operating expenses. In other words, total capital invested (K) equals the total cost (TC) of production.

Let's return to the initial position and now assume that the producer introduces a new technology which increases the total output with given inputs. In other words, all inputs used and the laborers employed have not changed but the quantity of output increased, due to the new technology. We observe once again that the added-value (VA), total profit (π) and the rate of profit (r) increase.

$$Q_{t+1} > Q_t$$
$$VA_{t+1} > VA_t$$
$$TC_{t+1} = TC_t \text{ and}$$
$$r_{t+1} > r_t$$

A new technology introducing a new production process may save some inputs or increase the supplied quantity or may cause the both at the same time. In other words, the quantity supplied may increase while the inputs used decrease due to the new technology. Or, the new technology may only save the laborer and requires fewer employees to produce the quantity of output. Table: 7-2 shows us how the total output (Q), the rate of profit (r), the value-added (VA= π +w*L), the partial factor productivity (PFP), total factor productivity (TFP), and the labor-wage productivity (LWP) would be effected.

Table: 7-2 Effects of a new production method, given the product
(all input and output prices (p_{ij}), wage (w) and interest rate (i) is constant)

Type of technological change	K-saving	Impact on output	Total VA	VA/PFP	VA/L	VA/LWP	q/L	VA/K	Change in r and π
Input-saving	Yes	$Q_{t+1}=Q_t$	↑	↑	↑	↑	Constant	↑	↑
L-saving	Yes	$Q_{t+1}=Q_t$	↑	Constant	↑	↑	↑	↑	↑
Q-increasing	Yes	$Q_{t+1}>Q_t$	↑	↑	↑	↑	↑	↑	↑
Q-increasing L & X saving	Yes	$Q_{t+1}>Q_t$	↑	↑	↑	↑	↑	↑	↑

Note: Demand increases at the same rate as output.
$PFP = VA / X_i * p_i$ (exc. L) $K = p_i X_i + wL$
$VA = w*L + \pi$ (incl. İnterest rate) $r = \pi / K$
$\pi = VA - w*L$ or TR - TC

Technological Productivity Growth

In the following parts, the purpose will be to display in simple models how technological productivity growth affects price, value-added, wage-profit ratio and functional income distribution.

"Given" Product – "New" Production Process

All initial production figures used are selected randomly. Since the sole purpose is to analyze the consequences of a technological innovation, the selection of random figures is unimportant.
  Initial values:

$w_t = 100$ TL
$L_t = 500$ workers
$LWC_t = w_t * L_t = 100*500$ = 50,000 TL
$OC_t = FC_t + VC_t = 40,000 + 40,000$ = 80,000 TL
$TC_t = LWC_t + OC_t$ = 130,000 TL

TC denotes total costs, OC total costs excluding wages, FC fixed-costs like rent-capital goods, VC variable costs excluding wages such as raw-materials, energy, etc., and t, the time. On the income side, (TR) denotes the total revenue.

$p_t = 15$ TL
$q_t = 10,000$ pieces
$TR_t = p_t (15) * q_t (10,000) = 150,000$ TL

p denotes price and q the quantity produced. The hypothetical size of profit (π), rate of profit (r) and value-added (VA) will emerge as follows:

$\pi_t = TR_t - TC_t$ = 20,000 TL
$VA_t = \pi_t + LWC_t = 20{,}000 + 50{,}000 = 70{,}000$ TL
$r_t = \pi_t / TC_t = 20{,}000 / 130{,}000 = \sim \%\,15$
$\pi_t / VA_t = \sim \%\,28$ (share of profit in VA)
$LWC_t / VA_t = \sim \%\,71$ (share of wages in VA)

Case-1 Laborer Saving Technological Innovation

Assume that after the introduction of new technology the firm continues to use the same quantities of inputs as before. But the number of employees is reduced from 500 to 300, cet. par. The new technology will alter the rate of profit and shares of income in VA of wage and profit, assuming that quantity supplied, price, wage-rate and non-wage input costs remain the same.

$LWC_{t+1} = w_{t+1} * L_{t+1} = 100 * 300$ = 30,000 TL
$OC_{t+1} = FC_{t+1} + VC_{t+1}$ = 80,000 TL
$TC_{t+1} = LWC_{t+1} + OC_{t+1}$ = 110,000 TL
$\pi_{t+1} = TR_{t+1} - TC_{t+1}$ = 40,000 TL
$VA_{t+1} = \pi_{t+1} + LWC_{t+1}$ = 40,000 + 30,000 = 70,000 TL

and;

$r_{t+1} = \pi_{t+1} / TC_{t+1} = \sim \%\,36$
$\pi_{t+1} / VA_{t+1} = \sim \%\,57$ (share of profit in VA)
$LWC_{t+1} / VA_{t+1} = \sim \%\,43$ (share of wages in VA)

Although the total VA produced is unchanged (150,000 TL), share of profit in VA (r/VA) increased from 28 percent to 57 percent. Although the real wage level remained the same as before, the share of wages in VA dropped as a consequence of laborer saving technological innovation.

Another interesting development is the decline in VA produced. The reason for this outcome is the reduction in the number of employees, which caused a decline in the total VA ($LWC = \pi$) produced.

By assumption, price was given and remained unchanged. As a result of new cost reducing technological innovation, the firm faces with three options:

1. Keeping the price intact and increase total profits.
2. Increase its competitive strength by reducing price, while keeping the profit rate constant.
3. A combination of options 1 and 2.

If the firm prefers to reduce the end price of product, it will gain competitive advantage but at the cost of losing some profits.

Case-2 Input-saving (Excluding Laborer) Technological Innovation

Let us assume that, after the introduction of new technology, the same quantity of output is produced with less amount of inputs but the same amount of laborer, say $VC_{t+1}=20,000$. Some variables will change, while others remain the same. Those remaining the same are:

$w_t = 100$ TL
$L_t = 500$ laborers
$p = 15$ TL
$q = 10,000$ pieces
$LWC_t = w_t * L_t = 100*500 = 50,000$ TL

And the new values of some variables:

$OC_{t+1} = FC_{t+1} + VC_{t+1} = 40,000 + 20,000 \quad = 60,000$ TL
$TC_{t+1} = LWC_{t+1} + OC_{t+1} = 50,000 + 60,000 \quad = 110,000$ TL
$\pi_{t+1} = TR_{t+1} - TC_{t+1} = 150,000 - 110,000 \quad = 40,000$ TL
$VA_{t+1} = \pi_{t+1} + LWC_{t+1} = 40,000 + 50,000 \quad = 90,000$ TL
$r_{t+1} = \pi_{t+1} / TC_{t+1} = 40,000 / 110,000 = \quad \sim \% 36$
$\pi_{t+1} / VA_{t+1} \quad = \quad \sim \% 44$ (share of profit in VA)
$LWC_{t+1} / VA_{t+1} = \quad \sim \% 55$ (share of wages in VA)

As displayed above, after the introduction of new input-saving technology the rate of profit increases from 15 percent to 36 percent, share of profit in VA from 28 percent to 44.4 percent, while share of wages in VA drop from 71 percent to 55.5 percent, cet. par.

Since, by assumption, there is no change in quantity demanded, there is no reason to reduce the price. However, the firm might desire to use the cost advantage in competition to expand its markets. In this case, the price will have to fall. Assume that initially the rate of profit remains the same ($r = \% 15$) and price is subject to fluctuation. Reducing the price from 15 TL to 12.70 TL would not cause any inconvenience to the firm. But, at any price between 15 TL and 12.7 TL would imply loss of potential profits, cet. par. Any amount of fall in price below 15 TL would also imply decline in the potential share of profits in VA.

P_{t+1} $\quad = \quad$ 12.70 TL
$TR_{t+1} = p_{t+1} (12.7) * q_{t+1} (10,000)$ $\quad = \quad$ 127,000 TL
$\pi_{t+1} = TR_{t+1} - TC_{t+1} = 127,000 - 110,000$ $\quad = \quad$ 17,000 TL
$VA_{t+1} = \pi_{t+1} + LWC_{t+1} = 17,000 / 50,000$ $\quad = \quad$ 67,000 TL

$r_{t+1} = \pi_{t+1} / TC_{t+1} = 17{,}000 / 110{,}000 =$ \qquad ~ % 15
$\pi_{t+1} / VA_{t+1} = 17{,}000 / 67{,}000 =$ \qquad ~ % 25 (share of profit in VA)
$LWC_{t+1}/VA_{t+1} = 50{,}000/67{,}000 =$ \qquad ~ % 75 (share of wages in VA)

Case-3 Output Increasing Technological Innovation

Now assume that all inputs of production, including the employees, remains the same while output increases by 20 percent from 10,000 to 12,000 due to new production process. Once again, some variables will change while others remain unchanged. First the constants:

$w_t = 100$ TL
$L_t = 500$ employees
$LWC_t = w_t * L_t = 100*500 = 50{,}000$ TL
$p = 15$ TL
$OC_t = FC_t + VC_t = 40{,}000 + 40{,}000 = 80{,}000$ TL
$TC_t = LWC_t + OC_t = 130{,}000$ TL

And the changed new values:

$q_{t+1} = 12{,}000$ pcs
$TR_{t+1} = p_{t+1} (15) * q_{t+1} (12{,}000)$ \qquad = 180,000 TL
$\pi_{t+1} = TR_{t+1} - TC_{t+1} = 180{,}000 - 130{,}000$ = 50,000 TL
$VA_{t+1} = \pi_{t+1} + LWC_{t+1} = 50{,}000 + 50{,}000$ = 100,000 TL
$r_{t+1} = \pi_{t+1} / TC_{t+1} = 50{,}000 / 130{,}000$ \qquad = ~ % 38
π_{t+1} / VA_{t+1} \qquad = % 50 \qquad (share of profit in VA)
LWC_{t+1} / VA_{t+1} \qquad = % 50 \qquad (share of wages in VA)

The new technology causes the initial profit rate of 15 percent to rise to about 38 percent, the share of profit in VA from 25 to 50 percent, while the share of real wage in VA drops from 75 percent to 50 percent, though it remained unchanged. Note that tough the real wage does not change, its share drops, implying improved distribution of functional income in favor of the profits due to new technology.

Assume that the firm decides to reduce the price of product from 15 TL to 13 TL to gain competitive price advantage. This move would cause the profits to fall below the potential amount, but make the firm more competitive.

$P_{t+1} = 13$ TL
$TR_{t+1} = p_{t+1} (13) * q_{t+1} (12{,}000)$ \qquad = 156,000 TL
$\pi_{t+1} = TR_{t+1} - TC_{t+1} = 156{,}000 - 130{,}000$ = 26,000 TL
$VA_{t+1} = \pi_{t+1} + LWC_{t+1} = 26{,}000 + 50{,}000$ = 76,000 TL
$r_{t+1} = \pi_{t+1} / TC_{t+1} = 26{,}000 / 130{,}000$ \qquad = % 20

$\pi_{t+1}/VA_{t+1} = 26{,}000 / 76{,}000 \quad = \sim \%\ 34$ (share of profit in VA)
$LWC_{t+1}/VA_{t+1} = 50{,}000/76{,}000 \quad = \sim \%\ 65$ (share of wages in VA)

In the case presented above, in spite of the price fall to 13 TL, the rate of profit is higher than in the "initial" case. In addition, the firm is in a better position in terms of share of income. The new technology has not only strengthened the competitive position of firm, but also raised both, the profit rate and share of profits in total revenue, in comparison to "initial" case.

Case-4 Labor- and Input-saving Technological Innovation

In actual economies, new technologies do not save "only" laborer or inputs, as discussed in the previous samples. "Given" the product, the new technology often saves both, laborer and inputs and occasionally even the output as well. Such developments cause the profit rate to rise, which is the main goal of every firm in the long-run. As a result, though the real wage rate does not change as in the previous samples, the share of profits in VA increases and the functional income is re-distributed in favor of the profits, cet. par.

Assume that the initial values are valid and new technology allows the firm to produce the same amount of output with less inputs and employees.

$w_{t+1} = 100$ TL
$L_{t+1} = 400$ employees
$p_{t+1} = 15$ TL / pcs
$q_{t+1} = 10{,}000$ units
$LWC_{t+1} = w_t * L_t = 40{,}000$ TL
$FC_{t+1} = 30{,}000$ TL
$VC_{t+1} = 30{,}000$ TL
$OC_{t+1} = FC_{t+1} + VC_{t+1}$ $\quad = 60{,}000$ TL
$TR_{t+1} = p_t*q_t = 15 * 10000$ $\quad = 150{,}000$ TL
$TC_{t+1} = OC_{t+1} + LWC_{t+1} = 60{,}000 + 40{,}000 = 100{,}000$ TL
$\pi_{t+1} = TR_{t+1} - TC_{t+1} = 150{,}000 - 100{,}000 \quad = 50{,}000$ TL
$r_{t+1} = \pi_{t+1} / TC_{t+1} = 50{,}000 / 100{,}000 \quad = \%\ 50$
$VA_{t+1} = LWC_{t+1} + \pi_{t+1} = 40{,}000 + 50{,}000 \quad = 90{,}000$ TL
$VA_{t+1} / L_{t+1} = 90{,}000 / 400 \quad = \quad 225$ TL / per employee
$\pi_{t+1} / VA_{t+1} = 50{,}000 / 90{,}000 \quad = \sim \%\ 55$ (share of profit in VA)
$LWC_{t+1} / VA_{t+1} = 40{,}000 / 90{,}000 \quad = \sim \%\ 44$ (share of wages in VA)

In consequence, as the costs of inputs and wages are reduced due to new technology, total profits (π_{t+1}) and profit rate (r_{t+1}) increases along with the share of profits in VA. Functional income is again re-distributed in favor of the profits while the real wage rate remains the same.

New Product

So far, we analyzed cases where a "given product" was produced by "new production processes". Yet, we know that productivity growth due application of new process technologies for "given products" has only a limited impact on the economic growth. For, no matter how much the production costs and/or price declines due to new process technology, the markets are bound to saturate sooner or later, which would bring the growth process to an end.

The actual cause of long-run productivity and economic growth is technological innovations introducing "new products", which are, in general, accompanied by "new production processes". Incessant supplies of new products are the cause of ever rising living standards. Due to such innovations, both the quality and the quantity of products placed at the service of end-users increases along with the individual and total wealth. In the absence of such innovations the marginal utility of the "given products" would tend to decline in time accompanied by declining profit rates. Eventually the economies would reach the well-known "equilibrium" point of Neoclassical doctrine at some future date and further growth would be subject to population growth rate only. New investments would cease and the living standards would remain unchanged. But this is not case in reality simply because of *new products/production processes*, which are the products of *creative mental labor*.

Let us take a closer look at how "new products/new production processes" affect the productivity change in terms of "aggregate value added" (AVA). Assume that in an economy five different products ($q_1, q_2, q_3, q_4,$ and q_5) are being produced. Q denotes the quantity of total output, Y the value of total output and p, the price of products.

$q_1 = 3,000$, $q_2 = 14,0000$, $q_3 = 5,000$, $q_4 = 7,000$, $q_5 = 6,000$ pieces
$p = 10$ TL
$Q_t = \Sigma\, q_i = 35,000$ *pieces* $i = 1,\ldots,5$
$Y_t = Q_1 * p = 35,000 * 10 = 350,000$ TL

Further assume that two "new products" are introduced following a technological innovations, say digital TV, q_6 and solar energy drive car, q_7. Total supply of new products is 5,000 pieces and the average price 10 TL, cet. par. Both, Q and Y will naturally change.

$q_6 = 2,000$ pieces,   and $q_7 = 3,000$ pieces
$Q_{t+1} = \Sigma\, q_i = 35,000 + 5,000 = 40,000$ *pieces*. $i = 1,\ldots.7$
$Y_{t+1} = Q_2 * p = 350,000 + 5,000 * 10 = 400,000$ TL

Thus, the contribution of new products to total wealth in terms of value will be $\Delta Y = 50{,}000$ TL. Meanwhile, total quantity supplied (Q) increases from 35,000 pieces to 40,000 and the total value-added (Y) from 350,000 TL to 400,000 TL.

The related and rather important questions are: How to determine the prices of new products? What would be the new profit-rate?

Since the products are *new ones*, there will be no chance to make a price comparison. But, since the owner of the new product will have monopoly rights due to patent ownership, the *"expected"* profit rate would, quite likely, be above the market average rate. This expectation is rather important for the further development and introduction of new products.

New Product and Functional Income Distribution

By assumption, the wage rate of an employee remained unchanged until the next round of wage-negotiations, while new technologies were being employed. Using simple mathematical symbols[20] the likely impact of technological innovation in short-run would be as indicated below, cet. par.:

$$w_{t+1} = w_t$$

But,

$$VA_{t+1} > VA_t$$
$$r_{t+1} > r_t$$
$$\pi_{t+1} > \pi_t$$
$$\pi_{t+1} / VA_{t+1} > \pi_t / VA_t$$
$$w_{t+1} / VA_{t+1} < w_t / VA_t$$

As observed above, the total value-added or total income increases with the employment of new technology, and the functional income changes in favor of the profits although there has been no decline in real wage-rate. In other words, every technological productivity growth implies re-distribution of functional income in favor of capital-owner, cet. par.

Wage rise

Deteriorated functional income distribution for wage-earners, in spite of constant real wage-level, as a consequence of technological productivity growth continues, normally, until the next round of wage-negotiations. The outcome of negotiations is uncertain and depends on the bargaining-power of both parties. Assume that, after the employment of new technology, the new values of certain variables emerge as follows:

---

20 w = wage; VA = value-added; t = time; r = profit rate; π = amount of profit

$w_t = 100$ TL
$L_t = 500$
$p_t = 15$ TL
$q_t = 10,000$ pieces
$LWC_t = w_t * L_t = 100*500 \quad = 50,000$ TL
$OC_{t+1} = FC_{t+1} + VC_{t+1} \quad\quad = 80,000$ TL
$TC_t = LWC_t + OC_t \quad\quad\quad = 130,000$ TL
$TR_t = 15 * 10,000 \quad\quad\quad\quad = 150,000$ TL
$\pi_t = TR_t - TC_t \quad\quad\quad\quad = 20,000$ TL
$r_t = \pi_t / TC_t \quad\quad\quad\quad\quad = \sim \% 15$
$VA_t = \pi_t + LWC_t = 20,000 + 50,000 = 70,000$ TL
$\pi_t / VA_t \quad\quad\quad\quad\quad\quad = \sim \% 28.5$
$LWC_t / VA_t \quad\quad\quad\quad\quad = \sim \% 71.4$

And further assume that after negotiations the wage-level increases by 20 percent.

$\Delta w = 20$

New wage level;

$w_{t+1} = 120$ TL

Naturally, both labor costs and total costs will rise.

$LWC_{t+1} = w_{t+1} * L_{t+1} = 120*500 \quad = 60,000$ TL
$TC_{t+1} = LWC_{t+1} + OC_{t+1} \quad\quad\quad = 140,000$ TL

And,

$\pi_{t+1} = TR_{t+1} - TC_{t+1} \quad\quad\quad = 10,000$ TL
$r_{t+1} = \pi_{t+1} / TC_{t+1} \quad\quad\quad\quad = \sim \% 7$
$VA_{t+1} = \pi_{t+1} + LWC_{t+1} = 10,000 + 60,000 = 70,000$ TL
$\pi_{t+1} / VA_{t+1} \quad\quad\quad\quad\quad = \sim \% 14$
$LWC_{t+1} / VA_{t+1} \quad\quad\quad\quad = \sim \% 85$

As a result of wage-rise, though the total value-added remains unchanged (70,000 TL), the share of profit in total value-added declines from 15 percent to 7 percent, while the share of wages increases from 71.4 percent to about 85 percent. The shares of wage-rate and profit-rate affect each other in the opposite directions.

New Product, "Monopoly" and Profit Rate

Assume that Company-X, using its mental endowments and accumulated knowledge, develops a new medicine against cancer and acquires the patent for exclusive use of it. Due to patent ownership, Company-X enjoys a rather privileged monopoly position in the market against competitors. In the initial stage of new product demand would quite likely by far exceed the supply and being a monopoly, Company-X would set the price as high as market can bear. Accordingly, the profit-rate would quite likely be above the market average rate, say by 50 percent. But in time, the competitors would develop similar products and as the supply increases, competition will force the price and the profit-rate to decline. Assume that, given time, the profit-rate falls to the market average-rate, say 10 percent. This hypothetical trend of profit-rate is illustrated on Figure: 7-3 which indicates that at time-period ($_{t+4}$) the profit-rate declines to average market profit-rate. In the absence of competition, Company-X would set the price as high as possible and extract above average profit-rate, as long as demand lasted.

Figure: 7-3   Competition and changing trend of profit-rate

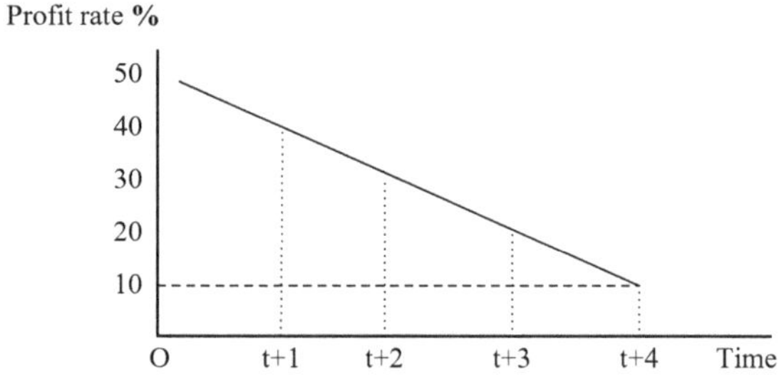

Innovations, Growth and Price

A "Given Product" but a "New Production Process"

Let us assume that the new technology is a new process technology enabling a decline in the production cost of a "given" product. For instance, say that the new process technology reduces the cost of refrigerator production from 100 TL to 90 TL per piece. This situation would give the producer three options regarding the price:

1- Continue to sell at the same price as before, while increasing the profit-rate:
2- Reduce the end-price to gain competitive edge against the competitors, cet. par.:
Or,
3- A combination of 1 and 2.

If first option is preferred, the price will remain unchanged, but the value-added created per unit-time employed and profit-rate would increase. If second option is preferred, both the price and the "potential" profit-rate would have to decline, while total revenue is likely to increase, cet. par. Third option would provide a combination of two outcomes.

A producer in a fair competitive environment with access to a new process technology enabling reduction in unit costs is quite likely to prefer the second option and reduce the end-price. This would be a rather rational behavior paving the way to a larger market share of the market and eventually to elimination of competition in the long-run, cet. par. The price reduction due to new process technology would also be in the favor of end-users, which would imply a positive "income effect" on their income.

"New" Product and "New" Price

Studying the long-run trends in economic development, we observe that the real price-level does not show continues tendency to decline. The main reason for this is the incessant introduction of "new products/processes", which does not only cause the profit-rate to rise above the average but also requires setting new prices for new products. The profit-rate of new product was expected to be higher than average because of monopoly position and excessive demand. As long as monopoly privileges through exclusive patent right prevails, monopoly profits may continue. But, in time, the competitors are expected to catch-up, increase the supply and force the price/profit level to fall.

Studies on long-run price indexes have to take into consideration that there are incessant introduction of new products/processes accompanied by "new prices". For instance, mobile phones supply, in practice, the same kind of service as the traditional house phones. But, nevertheless, they are no longer the same, because they contain rather different features. Therefore, more expensive mobile phones than traditional ones do not imply that the "general phone price level" has gone up. It would be a great error to apply one price index for all phones, for all phones are not the same quality. The same conditions apply for the car industry as well. 2004 year car model of a company is usually not the same in 2005 or in 2006 with respect to its features and price. Because of the differences in quality, a price comparison would be not only irrational but also erroneous. Due to the above mentioned reasons, there seems to be no price index available to measure the long-run price-level properly.

Other Factors Influencing Price Level

Imperfect Competition

If there was a perfectly competitive market, as the orthodox doctrines claim, a cost reducing technological innovation of a "given product" would automatically cause the price of product to fall, cet. par. Actually, some prices tend to decline from time to time as a result of cost-reducing technological innovations and competitive environment. For instance, in spite of various kinds of qualitative improvements, the prices of computers have, in general, fallen sharply in last decades. But, there are also some factors making a price-reduction undesirable for producers. The main factor is "imperfect competitive market conditions". If the market is dominated by a few "oligopolistic" companies or by a "monopolistic" seller, cost-reducing technological innovations may not reduce the expected results. The single seller may decide or a few sellers might agree on a non-price competition strategy, which would be in favor of growing profit-rate per unit output, but against the interests of end-users.

Wage Negotiations
Another factor, which might prevent a plausible price-reduction, might be the wage-increase demands of labor unions. It is not an infrequently encountered incidence that the company management meets the wage-increase demands with positive approach in order to prevent any possible conflict, which might cause trouble in supply. The higher the expectations of management, the greater might be the tendency to accept wage-increase demands. Otherwise, a conflict with the union may cause serious losses of markets and profits to competitors. Especially in cases of monopoly or oligopolistic competition, the management might be more inclined to have a positive approach to wage-rise, because they can easily manipulate the end-price, thus their profits, cet. par.

With respect to general price level, a wage-increase in a sector would imply, in a way, a general level price-reduction, that is an increase in income, cet. par. But this development might lead to deterioration in real incomes of wage-earners in other sectors, if the wage-rise leads to a general price-level increase, cet. par.

Services Sector
The wage-increase demands of labor unions does not only influence the price level in related sector where new technologies lead to cost-reduction, but also the wage and price levels in other sectors, though there may not be any technological innovations, thus productivity growth. For instance, a wage-rise in an industrial sector due to technological productivity growth may cause, indirectly, wage-increases in other sectors, such as service or public sector, which display, owing to their internal dynamics, relatively less productivity growth. Public sector services may prove to be a proper example. Although the productivity growth seems to be rather insignificant, if not non-existent, this does not stop the public servants to demand wage increases. As a result, though the economy may

experience decline in price in certain sectors due to technological productivity growth, the general price-level may display a rather different trend, cet. par.

Debt-Interest Payments
Another reason making the decline in general-price level an unattractive option is the total debts and interest payments of producers. If the price-level falls in a sector due to a technological innovation, this might put the producers, who borrowed to undertake production, to an awkward financial position. For the price-decline would imply decreased real profits and increased financial production costs with regard to amount borrowed and interests to be paid. Therefore, price-falls are not always desirable from the point of borrowers, cet. par.

Institutional / Cultural Infrastructure
Factors like bad governance of national economy, inappropriate interventions, and inadequate institutional and/or cultural infrastructure can also influence the general price level.

Technology Transfer & Long-run Growth

Development implies continuous and sustained economic growth in terms of per capita production in addition to social, cultural, political, institutional changes, whatever the yardstick of measurement is. An economy can grow through various ways, one of which, perhaps the healthiest way, is through the technological progress. Empirical evidence indicates that the fastest growing enterprises are the ones making best use of technological progress. That is because; technology is the basis of economic growth and is expected to pave the way to economic development. The gap between the developed and developing countries could be narrowed if the technology could be transferred through proper channels to the less developed countries.

Regarding the transfer of technology to developing countries, the degree of success depends partly on the nature of transactions and partly on the technological absorptive capacity of the recipient country. The outcome of the former is mainly influenced by the negotiations between the parties involved. But the latter, the absorptive capacity of the recipient country, prerequisites huge investments in human capital and infrastructure as well as long-term consistent economic policies. Because;

> *"... a society's capacity to adapt itself to the requisites of advanced technology and to adapt the advanced technology to its own circumstances and objectives, as well as its capacity to innovate, will depend in part on the intellectual skills, the acquired knowledge and*

know-how, the problem solving competencies -in a word, on the cognition possessed by those who constitute that society." (Solo;1966)

As Singer pointed out;

"... a country which has no national capacity cannot know what technology is available to be imported, what the most suitable technology for itself is, where the best sources for such technology are, and what the best forms are in which such technology should be embodied -let alone bargain effectively about the terms on which such imports take place." (Meier;1976;399).

As a result the transaction in technology may produce many undesired impacts on issues such as employment, foreign exchange, income distribution, etc.

Regarding the inadequacy of essential factors of development such as physical/financial capital, foreign currency, infrastructure, high-level man-power, etc., Foreign Direct Investments (FDIs) were expected to fill in the gap, to some extent, at least, and make significant contributions to the development of national economies of developing countries. But it was soon to be discovered that the technology market contained many imperfections. Various factors were restricting the full-utilization of the transferred technology in developing countries. Besides its indisputable benefits to the importing country, the FDI had some adverse impacts on important issues such as income distribution, employment generation and foreign currency reserves. There were even claims that the costs by far outweighed the benefits generated by foreign investment or even the *"non-transfer"* of technology implying that the host country had no or rather limited positive benefits in terms of having access to the advanced knowledge in spite of far reaching incentives and facilities provided for the foreign investors.

Measures to be Taken

In the field of technology, there are many problems facing the developing countries. Technology related problems rank among the most urgent problems. The technological, organizational and other related problems of under-development can be counter-acted by three major means;

1- By promoting and encouraging the development of indigenous technologies through various measures and incentives, in accordance with the available human and physical resources.
2- By transferring advanced technology through "appropriate" patent/license arrangements.
3- By transferring the technology through the foreign direct investments of globally operating enterprises.

Development of indigenous technologies to higher levels than those prevailing in developing countries is a rather costly and time-consuming and complex process. Tying the scarce recourses to the development of something that already exists (e.g. more advanced technology) may seem as an irrational attitude, misallocation of scarce resources. The common analogy in relation to the matter in question is: "Why discover America again?" Moreover, while scarce resources are tied up in a costly process outcome of which is uncertain, the industrialized countries may further widen the technological gap by introducing new and more advanced technologies. In short, there would be rather high degree of risk and uncertainty involved if the developing country scarce resources were to be devoted to indigenous technology development. One should note, however, that success is not a utopia. Japan, for example, is a late-comer of industrial development, and her success is largely the outcome of consistent economic policies encouraging indigenous resources to imitate and further develop the imported advanced technology through mainly patent and license arrangements. Country-specific conditions and the potential capabilities, both human and physical, of each country play a decisive role in the outcome of efforts.

Patent-right or license arrangements, as the means of technology transfer, provide the importing country with access to a "given" level of technology, subject to the clauses imposed by the seller. The contents of clauses (e.g. bargaining power of parties) determine the extent of costs and benefits accruing to the parties involved. If, for example, forward and backward linkages are restricted, say through grant-back clauses, the technology importing country's gain would be limited. Since the patent and/or license arrangements provide access to a certain level of knowledge, the imported technology is bound to be outmoded after some time, thus forcing the importer to resort to technology owner for a new arrangement, not to lag too far behind the up-to-date relevant technology.

The problem of underdevelopment is not only a technological problem, but it also reveals itself in terms of under-utilization of labor-force, lack of capital, scarcity of foreign exchange, etc. Private direct investments can (does) make valuable contributions to the relief of these problems in developing countries. Foreign direct investment by enterprises implies introduction of new products (product-diversification) and new production methods in the host-country. As a result of foreign direct investment, the domestic productive capacity and, most likely, the productivity increase along with the employment opportunities in the recipient country. The greater the total output, the greater would be the total income and income per capita. Since the foreign investors usually pay higher wages than the average, the income of the workers improve. Transfer of technical, organizational and managerial skills is another important contribution to the national economy of host country arising from the foreign direct investment. In addition, foreign investment saves, ceteris paribus, foreign currency by producing commodities that were previously being imported, and may even earn foreign currency through exports. In the absence of tax-holidays, foreign investor contributes to the revenue of state and municipalities as tax-payers. In view of all these "potential" advantages to be gained, ignoring the costs involved for the sake of argument, foreign direct investments by private enterprises in developing countries seem to be a valuable and indispensable source of economic growth. But, there are also costs involved.

Major Costs of Technology Transfer Through FDIs

The costs arising from the foreign direct investment can be studied in two major groups as;

1- Pre-investment (initial) costs; and
2- Post-investment costs.

Pre-investment costs cover all the expenses on pre-investment feasibility studies, management consultancy, construction of the administrative building and production plant, and other related expenditures. Depending on the nature of investment, at least some of the costs are financed by the host country shareholders. Pre-investment costs accruing to the host country constitute a relatively small portion of long-run total costs, and are easy to measure by the available data.

Post-investment costs, on the other hand, cover all the expenditures accruing to the host country in the period commencing from the operation of plant, and appear to burden solely the host country economy. The major post-investment costs can be classified in five groups:

1- Foreign exchange costs (remittances of profits, royalties, dividends, interest on foreign loans, salaries of foreign personnel, import of machinery, equipment and raw-materials).
2- Tax-concessions.
3- Economic dependence and vulnerability.
4- Inappropriate technology.
5- Restrictive clauses.

Foreign exchange costs and tax-concessions are statistical concepts and as such are easy to identify and measure in the national statistical accounts. Items 3 and 4, on the other hand, are analytical concepts and as such rather cumbersome to measure the extent of related costs. Costs arising from the restrictive practices of enterprises seem to be very serious ones. Contractual restrictive practices have been frequently used to control the quality, quantity, prices, exports, sources of imports, interest rate on loans, etc. Major consequent negative impacts of such practices can be summarized as follows:

1- Diffusion of transferred technological/managerial knowledge and skills is far from satisfactory (inadequate linkage effects);
2- Key decisions affecting the nature of transactions of the subsidiary and/or the interests of host country are often taken by foreigners;
3- Excessive foreign exchange costs as a result of *transfer pricing* mechanism;
4- Tax revenue losses arising from tax-incentives and undeclared profits (Transfer pricing);
5- Foreign exchange costs due to restrictive export clauses;
6- Economic/political dependence and vulnerability on external events;
7- Technological dependence perpetuated by market imperfections due to heavy reliance on foreign technology;
8- Pollution arising from the lack of effective protective measures and personnel to preserve the environment.

Conclusion

The main source of long-run economic growth, thus of ever rising standards of living, is technological progresses, which are a product of mental labor. Owing to technological progresses the long-run average profit-rate does not show a tendency to fall below the critical limits for continued new investments. Technological innovations do not only reduce production costs, but also introduce "new" products and processes, which are crucial for long-run growth trends.

The critical sort-term impacts of technological productivity growth, e.g., new technologies, are as follows, cet. par.:

a) Real wage rate remains unchanged while *VA* increases.
b) The relative share of real wage in *VA* declines.
c) The relative share of profits in *VA* increases.
d) As a consequence income distribution changes in favor of profits.
e) Income distribution trend is closely associated with real wage trend.

Functional income distribution changes in favor of profits, because the wage-rate is assumed to be constant in the sort-run until the next round of wage negotiations. But, in the course of time, as a consequence of positive changes due to technological productivity growth, demand and pressure for wage-rise increases to take a larger piece of economic pie produced. The final outcome depends on the bargaining power of two parties and on specific economic conditions.

"In the long-run", both individual and total wealth and standards of living tend to increase due to technological productivity growth. In other words, *incessant technological progresses appear as the source of long-run and sustained economic growth*.

Developed countries, which are quite well aware of the crucial role played by technological innovations, put great emphasis on education/training and R&D facilities. But, the situation in other countries lagging behind, like Turkey, Peru, Gambia, Vietnam, etc., is rather different. These countries do not necessarily need to make new inventions/innovations to secure growth, at least for a quite long period. Because, if the "available" technologies developed and owned, through patent agreements, by industrialized country producers can be transferred to developing countries through appropriate channels, they could make the same impact as new technologies do in developed parts of the world. Therefore, to benefit from technological innovations, developing country producers do not have to follow the highly costly and risky R&D process as developed country producers did. The basic pre-requirements for a successful transfer are access to, first of all, a labor force with appropriate qualities along with political-institutional and cultural infrastructure, at least "*in theory*".

But, unfortunately, the global markets for technology transfer are far from perfect (Gürak; 2003). Although the adaptive capacity of recipient country is rather important, that is not enough for a successful technology transfer. There is an urgent need to formulate new codes of global technology transfer in favor of the global competitiveness and, more importantly, in favor of the less developed countries.

Developing country decision-makers should also put great emphasis on increasing the qualities of their labor-force in order to make efficient use of available technologies and to successful adaptation/further development of new technologies to be transferred.

While studying long-run economic growth, technological progress and global economic relations, it seems rather important to attribute careful attention and to take the necessary precautions on the following critical aspects:

1. The characteristic features of global technology markets such as promotion of R&D, *ownership*, etc.
2. The process of technology transfer and the global technology market imperfections (Gürak, 2003).

3. Global distribution of global *VA* produced.

To Summarize;

- *Technological productivity growth* is the source of long-run economic growth.
- The source of technological productivity growth is *technological innovations*.
- And the source of all technological innovations is *mental labor*.

Efficient use of technologies requires *labor-force* with necessary qualitative endowments. And the long-run growth of economies requires laborer with *creative mental capabilities*.

Chapter-8
Growth & The Service Sector

Up to now we have used the word "product" to mean both physical and/or storable goods and non-physical products which cannot be stored; "services". This difference between physical and non-physical products is, generally, ignored in economic text-books. Traditional theories show clearly that what have actually been studied are the transactions concerning physical goods. As a result, the subject matter of these models has traditionally been the supply and demand for physical goods produced in the agricultural, industrial, mining or manufacturing sectors. This narrow focus has been continued up to now by many contemporary academics. Looking at today's "marginal productivity" analyses demonstrates this.

The difference between physical and non-physical goods should be self-evident. Physical products can be stored, transported and accumulated. *"Services" are products which are "consumed" right after the production process and cannot be stored, transported or accumulated.*

The services sector has been continuously growing in both the developed and developing countries. This is evident in terms of employment and the value added to the GDP. This means that the industrial and agricultural sectors have been declining in relative terms (see Table: 8-1). The relative proportion of Marx's "proletariat", i.e. "the blue-collar workers", has been declining while the proportion of the "white-collar workers" has been increasing "percentage-wise". The "proletariat" is being converted into, what Toffler called the *"cogniteria"*[21] (Toffler;1992; 90).

Table: 8-1 Proportion of the Service supply in the GDP (in %)

	1990	2003
Low income countries	41	49
Medium income countries	46	54
High income countries	65	71
World average	61	68

Source: World Development Report-2005, Table: 4-2

---

21 Cogniteria: Well educated, knowledable

It is neither sensible nor rational to continue to ignore the services sector in an economic theory. The growing significance of the services sector can only be ignored if one adheres strictly to some form of "ideological" model that ignores what is actually happening all around us.

The fact that the services sector has been increasing both in terms of employment and added value, does not mean that the importance of the industrial or manufacturing sectors is declining. On the contrary, there is a close correlation between more advanced physical products and the services sector. For example, as the quality of the airline service increases, so does the quality of the complementary services. As the quality of educational tools increase, so does the quality of the educators. As the quality of medical equipment increases so does the quality of the health services provided. In other words, there is direct correlation between industrial sector developments and the developments in the services sector.

"Productivity" & "Productivity Growth" in the Services Sector

Can we measure productivity in the services sector in the same way we do in the physical goods producing sector? Can we use the same criteria for both? What are the similarities or differences in these two sectors?

There are three main factors influencing productivity in the services sector:

1. The quality of the labor supplying the service;
2. The time spent on the supply of these services;
3. The level of technology employed by the service provider.

1- In the services sector, just as in the physical goods producing sector, there is a close correlation between the quality of labor and the level of productivity. If the quality of the labor is inadequate and inefficient, the quality of the service provided will be low.

2- The time-spent criterion is important in the analysis of services sector productivity, especially in comparative studies. But, whenever the quality of product enters the productivity analysis, using the "time-spent" measurement method becomes less effective, due to the difficulty in the measurement of quality.

3- The type of technology used is crucially important in any productivity analysis. For example, qualitative differences in the services provided by a "high-tech" hospital in comparison to one with only "low-tech" facilities will be significantly different. Given an appropriate quality of labor, the higher the level of the technology used in providing the service leads to a higher expectation in the quality of the service provided.

Quality is the most important component of a service. At this stage, technology is assumed to be a "given" and the quality of the service is at an optimum level. In the following sections, we will take a closer look at productivity in the services sector.

"Productivity "in the Services Sector

If "productivity" is defined as "the ratio of the output to the input of the "physical" quantities" (Q/I), we would be hard put to measure the productivity in services sector. Let's take a crude but apposite example: the common hair-cut. The basic *physical* inputs of the service can defined as; a comb, a pair of scissors, a chair, a mirror and the barber's labor. Assume that we can, somehow, calculate the quantity of the input per customer or per day. But how can we measure the output of the hair-cut service "quantitatively"? Since the service involves a "cutting of hair" should we use the quantity of hair actually cut in the measurement of productivity? Should we measure the quantity of the hair in length, volume or/ weight?

Could we possibly be rescued from this conundrum by using the "universally valid" neoclassical application of the superior "marginal productivity" analysis? Let's assume that a dedicated mainstream ideologue makes a brave attempt and claims that the "marginal productivity" notion can solve this matter. Then, the next crucial question would be; how does he actually "measure" the productivity of a comb, a pair of scissors, a mirror and a chair, i.e. the barber's physical inputs? Another crucial question; in what sense is a comb or a pair of scissors productive? What is it they produce, exactly?

If we leave aside the quantitative analysis of the services sector and use a "value" criterion, there is a totally different result. That is because; by measuring productivity in terms of the "value" would help to overcome the problems of a quantitative analysis. Firstly there would be no problem at all in measuring the cost of the input. The price of the hair-cut would be the value of the output. Thus, by subtracting the input cost the barber's income, we find the productivity of the barber in terms of the profit made. Output income divided by the input cost would give us the desired ratio.

$V^B = \pi$ = *Haircut income* minus *Input costs*

And the ratio of the output to the input:

V^B = *Haircut income / Input costs*

Or, alternatively, the ratio of profit would be;

$V^B = \pi$ / Input costs

V^B, designates barber's income; and π the profit.

"Productivity Growth" in the Services Sector

As we said before "productivity" and "productivity growth" have been used to mean the same thing. However there is a serious distinction between the two concepts. Productivity is a static (motionless) concept. "Productivity growth", on the other hand, refers to a dynamic process. A "quantitative" analysis cannot be applied to a dynamic growth process in the services sector. The most appropriate methods of measuring productivity growth appear to be;

1. the rate of growth in profit per unit of labor-time employed;
2. the rate of growth in the added-value (*VA*) per unit of labor-time employed.

The productivity growth equation for an enterprise.

$$g^f = (\pi_{t+1}/L_{t+1}) - (\pi_t/L_t) = \Delta\pi/L$$

The productivity growth equation for a national economy:

$$g = VA_{t+1} - VA_t = \Delta VA$$

If both the technology and the "type of service" are accepted as "givens", productivity may be increased by making some short-run improvements in "efficiency". This is subject to some constraints. For example, a barber working optimally and serving 10 customers a day cannot increase that number unless he/she works longer hours. However there are physical, biological and legal constraints with regard to working hours. These constraints apply to other professions as well e.g. teacher, doctor or a bus-driver. Therefore, even with the "given" technology there is a limit to the increase productivity due to the length of time one is able to work.

In the long run, productivity can be increased by introducing *new cost-saving technology*. For example, using cheaper inputs due to the new technology while providing the service, say in a hospital or in a barber's shop, would reduce the cost of production, cet. par. This in turn would lead to an increase in added value or profit.

"New" Types of Services and Long-run Growth

In a short term productivity growth analysis, of a "given type of service", the technology employed was considered as either a "given". In this case, growth is limited as the market saturates. For long-run and continuous growth in services sector, "new" types of services such as "space tours" have to be introduced.

Despite problems in defining and measuring productivity growth in the services sector, the World Bank attempted to measure productivity growth between 1980 and 2003. The World Bank while accepting the data was unreliable provided new data derived from GDP statistics. The data (Table: 8-2) shows a stable growth rate, in the same direction as GDP, in the services sector in "Low Income Countries" during this period.

Table: 8-2 The rates of growth of service output and the GDP (in %)

	Service	Output	GD	P
	1980-1990	1990-2003	1980-1990	1990-2003
Low income countries	5.1	5.9	4.4	4.7
Medium income countries	3.1	3.5	2.8	3.5
High income countries	3.4	3.1	3.4	2.6
World average	3.4	3.2	3.3	2.8

Source: World Development Report-2005, Table: 4-1

Competition in the Services Sector

When it comes to competition, the services sector has an important "edge" over the physical goods sector. The services sector is not bound by the patent laws and so it isn't affected by a period of "monopolistic power". For example a tourist facility or a health establishment cannot realize any monopolistic profits. Some services such as "consultation", "cleaning", and "entertainment" seem more prone to competition. The laws concerning "useful model" or "industrial design" are not considered to be on the same footing as the patent laws.

Competition conditions in some services sectors are not "perfect" nor does it appear "desirable". Some services sectors appear to mirror the operations of the physical goods producers. For example, the global cinema or TV-film distribution sector seems to be under the control of US-based enterprises. Global banking and financial transactions seem to be the domain of a few D.C. enterprises. And there are many more examples of this global domination.

The services sector is fundamentally important in terms of employment, national output and economic growth. But, this contribution is unfortunately ignored, both by the mainstream economic models and the relatively new endogenous economic growth models. Since Western economic models generally focus their studies, on supply and demand of physical goods, they are unable to or incapable of fully comprehending the actual short or long-run economic growth process. Economic theoreticians should give priority to introducing a new value-price theory for the services sector instead of introducing further unrealistic and misguided theories on growth.

Global (International) Trade in the Services Sector

General Agreement on Trade in Services (*GATS*) was signed, under the umbrella of World Trade Organization (*WTO*) to promote the trade in a wide range of services (construction, distribution, recreation, architecture, education, telecommunications tourism and transport, to name but a few.) These are the dynamic components in both a developed and a developing country's economy.

According to *GATS*, there four models in the supply of services:

1. *Cross-border trade*, such as communication and transport; a service supplied from a member country to another member country.
2. *Consumption abroad*, such as tourism and education; a service supplied from a member country to consumers in another member country.
3. *Commercial presence*, such as co-investments and representation of foreign banks and insurance companies; a service supplied by a member country firm through representation in another member country.
4. *Movement of natural persons*, such as consultancy and free circulation of individuals; a service supplied by a member country firm through individuals in another member country for a specific period.

The size of Global trade in services has been continuously increasing. Table: 8-3 shows the level of trade in low-, medium- and high-income countries in 2002. Not surprisingly, but unfortunately, the balance of trade is, once again, contrary to the interests of the low- and medium-income countries.

Table: 8-3 Global trade in services* (2002)

	Exports (million US $)	Imports (million US $)
Low income countries	40,966	52,561
Medium income countries	225,630	240,279
High income countries	1,244,630	1,182,565

Source: World Development Indicators, 2004; quoted from data on Table: 4-7 and Table: 4-8, p. 208 and 210

* Types of services: Transport, travel, insurance, finance, computers, information, communication and other commercial services.

In view of Table: 8-4, we observe that there is a huge gap between the incomes of Turkey (a medium income country) and the USA (a high income country). In 2001, Turkey has no royalty income while the royalty income of the USA is more than 38 billion dollars. Given the imperfections in the Global technology markets, it does not seem likely that this income gap will close. In all probability it will widen. This problem of technological development and related income differences seem to be valid in all low- and medium-income economies, and there seems to be no light at the end of the tunnel.

Table: 8-4 Comparison of some commercial service data Turkey and USA (2001)

	Turkey (million US $)	USA (million USA $)
Incomes:		
Insurance	25	15,210
Financial	331	5,140
Royalty-License fees	-	38,660
Other commercial incomes	2,856	49,670
Expenditures:		
Insurance	-282	-4,890
Financial	-722	-4,010
Royalty-License fees	-119	-16,360
Other commercial expenditure	-1,163	-37,530

Source: IMF, Balance of Payments Statistics Yearbook, 2002; quoted from the data on p. 907 and 940.

In the year 2001, Turkey has no royalty income while the income of the USA is more than 38 billion dollars. Given the Global technology market imperfections, it does not seem likely to close this income gap; in fact, it would not be irrational to expect the gap to widen. This problem of technological development and related income differences seem to be valid in all low- and medium-income economies, and there seems no light at the end of the tunnel.

Chapter-9
Growth & Income Distribution

Functional Income Distribution

Unequal distribution of income has always been one of the major problem areas and often a cause of embarrassment for both economic science and economists. The problem does not refer only to income distribution and the related inequalities among individuals, households, and classes but also to global income distribution and inequalities among nations. Present global income distribution is especially disappointing when one considers it in terms of the expectations from globalization.

Traditionally "leftist" economists have always tended to emphasize the evils of income inequalities and supported the redistribution of income in favor of less favored groups in society. The conservative right wing economists, on the other hand, tended to give priority to capital accumulation rather than fair income redistribution which they believed would enlarge the cake to be shared out. As a result, the low income groups were expected to be better off.

The critical question in regard to the long-run economic interests of society is whether priority should be given to more fair income redistribution or to capital accumulation. Perhaps a combination of two policies would have a more positive impact in the long-run.

Functional income distribution, on the other hand, has been the subject to endless arguments ever since the appearance of economics as a separate discipline. Functional income distribution studies the distribution of income among two basic classes; workers receiving wages (W) and capitalists receiving profits (π) which includes interest rate and rent.

"Capital" was defined as any kind of expenditure with the purpose of producing goods and/or services in order to make profits. It can be expressed in terms of money, but it is not purely money. It covers all expenditures on the inputs of production ranging from patents or licenses for technology, to raw materials, semi-finished products, and energy and wages for the laborers. A capitalist assumes that he/she may risk losing some or all of his/her expenditure. In return of the risk assumed, a capitalist expects to receive some profit. Except in the minds of economists of the neoclassical heritage, in reality there is no such thing as the "productivity of capital". It exists only in unfeasible world of utopian models, because the two sole "productive factors" are nature and labor-power. "Marginal productivity" is simply an ideological concept fabricated by the ideologues of main-stream theories.

An Ideal Income Distribution

Pareto Optimum

The Pareto optimum allocation of income implies that one cannot redistribute income to make at least one individual better off without making others worse off. To put it differently, if it is possible to improve an individual's income position without worsening someone else's position, distribution is not yet in Pareto optimum.

The Pareto optimality depends on above mentioned utopian equilibrium and perfect competition conditions as well as "given" values. Many economists see some merits in such a deficient and utopian theory to explain actual economic developments. Another serious shortcoming of the theory is the lack of awareness of the impact of technological progress on income distribution.

Figure: 9-1 shows with the assistance of the so-called "Edgeworth Box" where the indifference curves of two consumers (O_a and O_b) are tangential. The Figure shows that there is more than one Pareto optimum points such as D^1 and D^2. Thus, at any equilibrium point along the OaD line, one should not attempt to redistribute income. D, D^1 and D^2 denote the possible equilibrium points and TT the transformation curve.

In fact, it seems that the Pareto optimum argument as such appears to serve the interests of higher income groups rather than aiming at a more equal income distribution, given the present situation of the unequal distribution of income. Because, redistribution of income in favor of the poorer income groups would lead to a deterioration of the high income groups' incomes that means at least one well-off individual of the upper class income group will be worse-off after the redistribution. Therefore it is better not to change the present equilibrium condition which is at the Pareto optimum, though this is not an egalitarian state of affairs.

Figure: 9-1 Pareto optimum income distribution

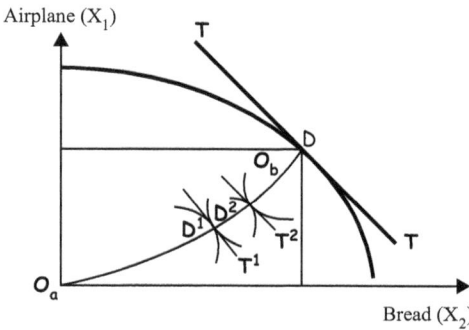

The functional income distribution we discuss in this section is in terms of the distribution of "added-value" which consists of both gross wages and gross profits. We shall first study a case to show how an efficiency increase (growth) effects the income distribution with a "given" technology in the short-run; followed by an analysis to show the effects of innovations on income distribution in the long-run.

Optimum Functional Income Distribution and Growth

Functional income consists of the wages (w) received by workers and the profits (π) of capitalists, including interest and rent incomes. In other words, it is the total "added-value" produced. However, it does not seem quite appropriate to consider rent and especially interest income as a sub-income of profits in contemporary income models. The reason behind this is the fact that a considerably large part of interest incomes do not come from loans for investment. There are huge amounts of financial resources looking for opportunities to exploit the position of governments or enterprises that are in financial straits. Sometimes, even enterprises use such opportunities to increase enterprise incomes. For instance, in years 2000 and 2001 a considerable part of the annual income of Turkish large-scale enterprises came from such financial transactions (see Table: 9-1). Thus, it would be an error to consider that all interest incomes come from loaned money used for investments. However, in this section, incomes from such financial transactions shall be ignored and adopt the position, like the Classical economists that functional income originates from two sources which are;

1- wages (w) received for the labor services hired; and
2- profits (π) received for the risks assumed.

The critical question here is; which kind of functional income distribution should be accepted as the optimum? In other words, what should be the optimum wages and profits?

The answer to this critical question is bound to be subjective and ideological, for it is impossible to provide an objective answer that is free of all value and ideological judgments on social, political or economic issues where two income receiving groups display diametrically opposite economic interests. One income group has to gain or lose at the cost of the other group. However, if capital ownership could be distributed equally amongst all individuals, the optimality problem would be quite different (Gürak, 2004).

Table: 9-1: Non-industrial Income of the Largest 500 Turkish Private Enterprises and the Ratio of Net Profits (before tax) (000 TL)

	Other incomes (1)	Change (%)	Net profits (2)	Change (%)	1:2 (%)
1982	18,438,880	-	120,558,000	-	15.3
1983	37,782,726	104.9	192,819,000	59.9	19.6
1984	65,888,349	74.4	316,573,000	64.2	20.8
1985	104,501,915	58.6	433,462,000	36.9	24.1
1986	204,558,231	95.7	663,879,000	53.2	30.8
1987	285,321,034	39.5	1,593,763,022	140.1	17.9
1988	632,251,158	121.6	2,494,087,226	56.5	25.4
1989	1,299,143,459	105.5	4,189,117,743	68.0	31.0
1990	2,224,648,247	71.2	6,679,368,414	59.4	33.3
1991	3,721,503,901	67.3	7,282,297,783	9.0	51.1
1992	7,794,368,122	169.7	20,052,422,789	175.4	38.9
1993	17,548,778,004	125.1	43,093,652,150	114.9	40.7
1994	57,694,648,710	228.8	105,587,224,479	145.0	54.6
1995	96,191,958,000	66.7	206,857,875,064	95.9	46.5

Source: İSO, Sanayi Odası Dergisi (Istanbul Chamber of Industry), Sept.-1996, No:366;82

In the following sub-sections the focus will be on the functional distribution of income amongst two classes; workers and capitalists. We shall first consider a short-run case to see how an improvement in "efficiency" with a "given" technology influences the functional income distribution. Then we shall attempt to show how innovations effect functional income distribution in the long-run.

The subjects to be discussed are:

1- Functional income distribution after efficiency increase, "given" a specific technology.
2- Impact of technological progress on functional income distribution.
3- Global functional income distribution.

Initial Case: Functional Income Distribution: with a "Given" Technology

We shall use a method based on the "added value produced" criterion in income distribution analysis. This approach is also used globally for national account statistics to measure growth and prosperity.

Assumptions:

- Prices for produced goods and services are a "given".
- Technology employed is a "given".
- No scarcity of endowed labor force required for efficient production.

Functional income in terms of the value added (VA) approach can be shown as below:

$$VA = W + \pi$$

where W stands for total wages and Π for total profits. The interests of these two income groups are diametrically opposed. If the share of wages in "added value" increases the share of profits is bound to decline and vice versa.

Figure: 9-2 shows this contradictory relationship between income classes at an enterprise level. Let us assume that the initial position is at B, the break-even point where costs equals enterprise income. Given the wage level and the price of the product, the area denoted by TR-TC-B indicates the total profits of the enterprise. If the wage level increases for some reason, cet. par., the total cost curve (TC) of the enterprise will shift further away from the origin and the new profit area will be smaller (TR-TC-B').

Figure: 9-2

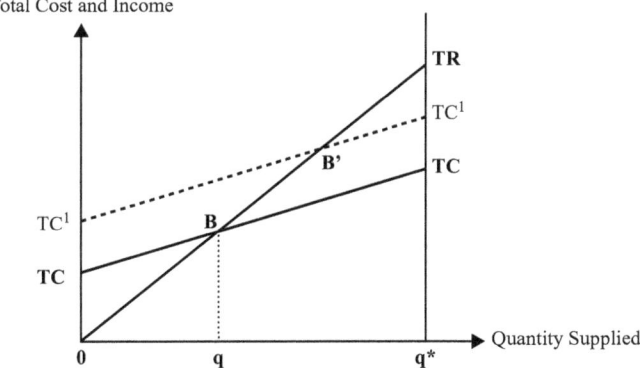

1- "Efficiency Growth" & Income Distribution

Given the wage level, the price of the product and the technology, and assume that an improvement is made in efficiency. As a result, the relative share of the wages in total added value would decline, while the share of the profits would increase leading to the laborers being relatively worse-off in terms of income (see Table: 9-2), cet. par. In other words, capital owners receive a larger slice of the pie while the laborers would have to settle for a smaller portion, cet. par.

Given time, the employees would, quite likely, attempt to increase their wage level which is dependent on their negotiation power and the market conditions. If wages increase with a given technology, i.e., method of production, the relative share of wages would improve but at the expense of the capitalists.

The nominal wage level is likely to increase in the long run as the economy grows. What happens to the real wage level is dependent, in the long-run, on the technological progress, the competitors' prices and the demand for the product along with the negotiating power of the trade unions.

Table: 9-2: Impacts of efficiency growth on income Price, technology and wage-level "given"

Cause of efficiency growth	VA:K	VA:L	π:VA	w:VA	Outcome
Restructuring production	↑	↑	↑	↓	Income distribution shifts in favor of capital-owner
Increasing capacity utilization	↑	↑	↑	↓	-- " --
Multiple shift work	↑	↑	↑	↓	-- " --
Reallocating resources	↑	↑	↑	↓	-- " --
Improved education system	↑	↑	↑	↓	-- " --
Learning By Doing	↑	↑	↑	↓	-- " --
Improved Safety & Sanitary	↑	↑	↑	↓	-- " --
Participation in decision-making	↑	↑	↑	↓	-- " --

VA: Value added. K: Total costs + wages, L: Labor-force; employees,.w: Wage.π: Profit

2- "Innovation" & Income Distribution

This section will analyze the impact of technological progress on functional income distribution. The main assumptions are as follows:

- Constant price.
- No decline in the effective demand.

Hypothetical initial values:

w_t = 100 TL
L_t = 500 employees
$LWC_t = w_t * L_t$ = 100*500 = 50,000 TL
$OC_t = FC_t + VC_t$ = 80,000 TL
$TC_t = LWC_t + OC_t$ = 50,000+80,000 = 130,000 TL

Total income (TR) of the enterprise:

p_t = 15 TL
q_t = 10,000 units
$TR_t = p_t (15) * q_t (10,000)$ = 150,000 TL

Profits (π), profit rate (r) and added value (VA) are.

$\pi_t = TR_t - TC_t$ = 150,000 - 130,000 = 20,000 TL
$r_t = \pi_t : TC_t$ = ~ % 15
$VA_t = \pi_t + LWC_t$ = 20,000+50,000 = 70,000 TL

And, the functional income distribution:

$\pi_t : VA_t$ = 20,000:70,000 =   ~% 28 (capitalist's share)
$LWC_t : VA_t$ = 50,000:70,000 =   ~%71 (workers' share)

In the analysis below, we shall see how technological progress affects functional income distribution. We shall study first the impact of laborer-saving then input-saving and finally output-increasing technological progress, in that order.

Case-1 Laborer-saving Technological Progress

Assume that the producing enterprise introduces a new production method which increases the productivity per employee by using labor saving inputs and the number of employees declines from 500 to 300. The price of product, wage and other input costs are assumed to be constant. What would be the impact of the technological progress on the functional income distribution?

$LWC_{t+1} = w_{t+1} * L_{t+1} = 100*300 = 30,000$ TL
$OC_{t+1} = FC_{t+1} + VC_{t+1} = 80,000$ TL
$TC_{t+1} = LWC_{t+1} + OC_{t+1} = 110,000$ TL

And,

$\pi_{t+1} = TR_{t+1} - TC_{t+1} = 40,000$ TL
$r_{t+1} = \pi_{t+1} : TC_{t+1} = \sim \% 36$
$VA_{t+1} = \pi_{t+1} + LWC_{t+1} = 40,000+30,000 = 70,000$ TL

New functional income distribution would look like:

$\pi_{t+1} : VA_{t+1} =$ $\sim \% 57$ (capitalist's share)
$LWC_{t+1} : VA_{t+1} = 30,000 : 70,000 =$ $\sim \% 43$ (workers' share)

Due to the decline in the number of employees who continue to produce the same amount of output, the percentage of profit in total added value ($\pi:VA$) increases from 37 percent to 57 percent, while that of the wage-earners' ($LWC:VA$) drops from 62 percent to 43 percent. Now, in terms of functional income, capital owners are better-off after the introduction of the technological progress while the workers are worse-off, cet. par. This outcome clearly shows why enterprises are in a constant search for new knowledge on production, i.e., new technologies, in order to reduce production costs.

Case-2 Input-saving Technological Progress
Assume that the initial values prevail and the enterprise introduces a new production method which reduces the quantities of non-labor inputs. Say that operating costs (OC) decline from 80,000 TL to 60,000 TL.

$w_t = w_{t+1} = 100$ TL
$L_t = L_{t+1} = 500$ employees
$LWC_t = LWC_{t+1} = 100*500 = 50,000$ TL
$OC_t = FC_t + VC_t = 80,000$ TL
$OC_{t+1} = FC_{t+1} + VC_{t+1} = 60,000$ TL
$TC_{t+1} = LWC_{t+1} + OC_{t+1} = 50,000+60,000 = 110,000$ TL

Total income (TR) of the enterprise is:

$p_t = p_{t+1} = 15$ TL
$q_t = q_{t+1} = 10,000$ pieces
$TR_t = TR_{t+1} = 150,000$ TL

New size of profits (π), profit rate (r) and added value (VA) would be:

$\pi_{t+1} = TR_{t+1} - TC_{t+1} = 150{,}000 - 110{,}000 = 40{,}000$ TL
$r_{t+1} = \pi_{t+1} : TC_{t+1} = \sim \% 36$
$VA_{t+1} = \pi_{t+1} + LWC_{t+1} = 40{,}000 + 50{,}000 = 90{,}000$ TL

And, inevitably, functional distribution of income would also be affected.

$\pi_{t+1} : VA_{t+1} = 40{,}000 : 90{,}000 =$ ~ % 44 (capitalist's share)
$LWC_{t+1} : VA_{t+1} = 50{,}000 : 90{,}000 =$ ~% 55 (workers' share)

As a result of the new technology, the capitalist's share in total added value ($\pi : VA$) increases from 28 percent to 44 percent, while wage-earners share ($LWC : VA$) declines from 71 percent to 55 percent. In the meantime, the profit rate jumps from 15 percent to 36 percent. We see clearly once again why enterprises are in a constant search for some new knowledge (technology) which reduces cost.

Case-3 Output Increasing Technological Progress

Assume once again that initial values prevail and the enterprise introduces a new production method which increases the quantities supplied with given inputs and number of employees. Say that the quantity supplied after the introduction of a new technology increases from 10,000 units to 12,000 units which inevitably would affect the distribution of the functional income.

$\Delta q = 2{,}000$ units
$q_t = 10{,}000$
$q_{t+1} = 12{,}000$ units
$OC_t = OC_{t+1} = 80{,}000$ TL
$TC_t = TC_{t+1} = = 130{,}000$ TL
$TR_t = 150{,}000$ TL
$TR_{t+1} = 15 \text{TL} * 12{,}000 = 180{,}000$ TL

Due to technological progress, capitalist's share in total added value will increase while that of wages decline, indicating a deterioration for the employees.

$\pi_{t+1} = TR_{t+1} - TC_{t+1} = 180{,}000 - 130{,}000 = 50{,}000$ TL
$r_{t+1} = \pi_{t+1} : TC_{t+1} = \sim \% 27$
$VA_{t+1} = \pi_{t+1} + LWC_{t+1} = 50{,}000 + 50{,}000 = 100{,}000$ TL

New functional income distribution would look like as follows:

$\pi_{t+1} : VA_{t+1} = 50{,}000 : 100{,}000 =$ ~ % 50 (capital's share)
$LWC_{t+1} : VA_{t+1} = 50{,}000 : 100{,}000 =$ ~ % 50 (workers' share)

Once again we observe that given the prices, wage level and demand, any technological progress produces results in favor of the capital owners, which is why enterprises constantly seek new technologies.

Wage Rise in the Long-run & its Impact on Income Distribution

In the analyses above we assumed a constant wage level in the short term until new wage negotiations between employee and employer were entered into. A constant wage level and technological progress seemed to erode the share of income of the employees even though the real wage may remain constant. When negotiations begin employee unions would normally demand a wage-rise while the employer unions attempt to resist this increase. The rise of the wage level depends largely on the negotiating power of the employee unions, the competitive conditions and the demand for products. Any improvement in favor of one income group would naturally imply deterioration in the relative share of the other group. However, wage negotiations, not infrequently, conclude with a percentage wage rise.

Assume that, as in *Case-1* above, the new technology reduces the number of employees and the labor-unions succeed in raising the wage level, say by 50 percent, cet. par.

$\Delta w = 50$ TL (wage-rise)

New wage:

$w_{t+2} = 150$ TL $(w_{t+1} + \Delta w)$

The wage cost of production would naturally increase and pull total costs upwards.

$LWC_{t+2} = w_{t+2} * L_{t+2} = 150*300 = 45,000$ TL
$OC_{t+2} = FC_{t+2} + VC_{t+2} = \qquad\qquad 80,000$ TL
$TC_{t+2} = LWC_{t+2} + OC_{t+2} = \qquad\quad 125,000$ TL

And,

$\pi_{t+2} = TR_{t+2} - TC_{t+2} = 25,000$ TL
$r_{t+2} = \pi_{t+2} : TC_{t+2} = \% 20$
$VA_{t+2} = \pi_{t+2} + LWC_{t+2} = 25,000+45,000 = 70,000$

New shares in total income:

$\pi_{t+2}\sim: VA_{t+2} = \qquad\qquad\qquad \sim \% 35$ *(capital's share)*
$LWC_{t+2}:VA_{t+2} = 45,000:70,000 = \sim \% 64$ *(workers' share)*

Given the size of profits and the total added value, the share of the capital in total added value declines from 60 percent to about 35 percent, simply because of a wage rise, while the share of the wage-earner increases to about 64 percent, cet. par. The wage rise clearly improves functional income distribution in favor of the working class and that is why enterprises always tend to resist wage rises.

Technological Imperfections & Global Income Distribution

It is not only the competition, productivity, prosperity and the income distribution of a country but also the global competition, productivity, prosperity and income distribution that are affected by technological developments. Per capita income as well as gross domestic product (GDP) is higher in high-technology countries compared to the technologically less advanced countries. For instance, Turkish enterprises, on average, use less advanced technologies in their production in comparison to Japanese enterprises. This is the underlying cause of lower per capita productivity and income in Turkey.

Let us analyze these global differences with the assistance of some hypothetical figures.

"Technology-intensive" & "Labor-intensive" Methods of Production

Two Products - Two Technologies - Two Countries
Assumptions:

- Two countries (the developed (X): the less developed (Y)
- Each country produces only one product.
- Developed country (X) employs advanced technology while the less developed country (Y) employs relatively less advanced technology.
- Due to technological difference, the wage level in the developed country (X) is, say four times, higher than in the less developed country (Y). In contrast, country (X) requires only one tenth of the employees compared to country (Y) to realize efficient production.
- In each country production is undertaken by one enterprise only with "exactly equal" sum of capital (K), say 100,000 TL.

Hypothetical values:

Developed Country (X) *Less Developed Country (Y)*
$w^x = 400$ $w^y = 100$
$L^x = 50$ employee $L^y = 500$ employee
$LWC^x = 20,000$ $LWC^y = 50,000$ ($LWC = w*L$)

$OC^x = 80{,}000$ $\qquad\qquad OC^y = 50{,}000$ $\qquad\qquad (OC = FC+VC)$
$K^x = TC^x = 100{,}000$ TL $\qquad K^y = TC^y = 100{,}000$ TL $\qquad (K = OC+LWC)$

Further assume that prices, quantities supplied and total incomes are as follows:

$p^x = 200$ TL $\qquad\qquad p^y = 150$ TL
$q^x = 1{,}000$ $\qquad\qquad\quad q^y = 1{,}000$
$TR^x = 200{,}000$ TL $\qquad TR^y = 150{,}000$ TL

It is highly likely that the higher percentage share and cost of the technologically advanced inputs of production would make the ratio of non-wage costs (OC) to total cost (TC) look higher in the developed country-X than in the less developed country-Y.

$OC^x: K^x > OC^y: K^y$

On the other hand, since the technology employed in the less developed country is relatively less advanced, the ratio of non-wage costs (OC) to the capital invested is only 50 percent.

Accordingly, the size of profits (π), rate of profits (r) and added value (VA) would look like as follows:

$\pi^x = TR^x - TC^x = 100{,}000$ TL $\qquad \pi^y = TR^y - TC^y = 50{,}000$ TL
$r^x = \pi^x : TC^x = \%\,100$ $\qquad\qquad\quad r^y = \pi^y : TC^y = \%\,50$
$VA^x = \pi^x + LWC^x = 120{,}000$ TL $\qquad VA^y = \pi^y + LWC^x = 100{,}000$ TL

And income distribution:

$\pi^x : VA^x \quad = \sim \%\,83 \qquad \pi^y : VA^y \quad = \%\,50$
$LWC^x_t : VA^x_t = \sim \%\,16 \qquad LWC^y_t : VA^y_t = \%\,50$

The hypothetical figures show that 100,000 TL invested as capital in high technology produces a value worth of 120,000 TL while the less advanced technology produces only 100,000 TL worth value. By the way, not only the added value but also the wage level and profitability are higher in the more advanced technology using country.

Given the insights above, in order to reduce and eventually eliminate the global income differences, it is imperative to eliminate technological differences in production and the imperfections in the global technology markets.

Concluding Remarks

In the short term, given the technology and the wage level, an efficiency growth is expected to increase the total added value, cet. par. while;

1- The real wage remains constant.
2- The proportionate share of wages decline.
3- The proportionate share of profits increase.

As a result, functional income distribution changes in favor of the capitalists and against the workers.

In the long run, total added value shows a tendency to rise alongside technological progress; a process called "technological productivity growth". Given the wage level, technological progress causes a change in functional income distribution in favor of the capitalist, not the workers, at the start of the process. However, in time, how the income or total amount of added value is distributed among the income groups depends on the bargaining power of the employee and employer unions along with the prevailing competitive conditions. Any change in any direction of functional income distribution leads to a deterioration in the income of the other group.

Chapter-10
Growth or Development?

Economy textbooks, in general, discuss the growth phenomenon in the light of full employment (the Neoclassical School) or underemployment (the Keynesian School) equilibrium models. In recent years, the *rediscovery*[22] of the "endogenous" growth models, have led to attempts to overcome a vicious circle which has proved fruitless and for many years lead to the presumption that technology is a "given" factor in economics. It is an obvious fact that, despite all the refinements in content or scope, the traditional or "utopian" academic models have not found any realistic solutions to the real economic growth problems of the developing countries. Therefore, as a necessity, in addition to these "utopian" academic growth models, a new and distinct study of economic analysis emerged under the name of the *"Development Economics"* which was used to analyze and solve the specific problems of the developing countries.

When one scrutinizes the academic economic textbooks, one concludes that the similarities and distinctions between growth and development have not been successfully addressed. According to some theorists, *growth* means an "increase in the total income of the developed countries", while *development* implies a "growth in the total income of the less-developed countries". Sometimes, in order to achieve development the emphasis is placed on a long-run successive production rise. For others, *growth* implies a long-term rise in per capita production or a rise in the production capacity at a level of full employment.

According to Todaro, a notable theoretician of Development Economics, mainstream neoclassical economy deals with the "perfect" markets of the advanced capitalist countries (Todaro;2003;8). However, neither the developing countries' markets nor their other institutions are perfect. On the contrary, their markets are imperfect and they suffer from other problems, be they economic or non-economic. Since the mainstream economic theories cannot help us in understanding and solving developing country problems, a *"different"* economic

---

22 A "rediscovery" since technological inventions are, according to the Classical economists of 150-200 years ago, particularly Karl Marx, the most integral features of the capitalist system, very far from "being exogenous or having unknown origin", Long before Schumpeter, Marx had emphasized why technological inventions are so important and their impacts across the economy. Thus, *"endogenous growth models"* embracing technological inventions are in fact not a novelty, rather, *rediscovered facts*.

approach seemed necessary. Therefore, for developing countries, a 'unique' economic approach was required, which must take into consideration not only the effective distribution of resources and sustainable development but also the social, political and institutional mechanisms and changes in those mechanisms of these countries. Therefore, changes in social indicators became as significant as the economic indicators.

Economic growth due to technological progress is often accompanied by institutional and cultural changes. For instance, in advanced countries, the Internet not only paved the way for new lines of business, but also contributed to issues such as increasing rates of social transparency, information exchange, as well as political participation. Therefore, institutional reorganization has taken place along with economic growth and socio-economic changes. We can also see similar changes in developing countries. If we consider that such changes, coupled up with technological inventions, somehow, constantly influence and change our lifestyles, we should be very careful about the line we draw between growth and development.

"*The multidimensional process of change*" as coined by Todaro (2003) for developing countries, in reality, is a process which also affects advanced countries. The difference is the desire of the developing countries to have a faster and greater rate of change. Otherwise, it would not be possible to close the gap. The World Bank, in its publication of the "Quality of Growth" (2000), points out that improvements in human, physical, environmental as well as the administrative aspects are required to ensure the quality of growth in the developing countries. Such transformations, along with technological inventions and growth, have also been taking place in the developed countries.

Growth

We defined the growth as the increase in the value added (*VA*) to a product.

$g = \Delta\ VA = VA_{t+1} - VA_t$

While it is possible to measure the rate of growth (g) according to various criteria, the least dubious measurement method is to measure the change in the value added (*VA*) to a product according to the total population (*N*), or per laborer (*L*), or per labor-hour (L_{hour}), or the labor wage cost (*LWC*) in the following manner.

$g = VA_{t+1} - VA_t\ /\ N$
$g = VA_{t+1} - VA_t\ /\ L$
$g = VA_{t+1} - VA_t\ /\ L_{hour}$

$$g = VA_{t+1} - VA_t / LWC$$

Regardless of which the above criteria is taken into account, we see that long-run growth is achieved in both advanced and developing countries. However, growth is realized at relatively higher rates in some countries and at relatively lower rates in others. Nevertheless, it is an undeniable fact that all countries achieve growth in the long-run.

The lesson to be drawn from this is that growth is a reality which is observed in advanced as well as developing countries[23].

Development

It is well-known, thanks to the contribution of Solow; the significance of the technological inventions has been "*re-discovered*" by the mainstream economists. But, a very serious problem emerged; the origin of these technological inventions essential for the growth process was "unknown". The growth models which contained technological inventions with "unknown" origin, did not pose a significant problem for the developed countries and their firms. That is because; introducing new technologies was an "internal" and "inevitable" process for the competing firms. Therefore, Solow's contribution was in fact more significant for the "academic" analysis rather than for the practicing businessmen.

When we evaluate these approaches in terms of the developing countries, we enter a completely different realm. Problems in the developing countries were incredibly complicated and intensive. Moreover in the "scientific" (?) theories, particularly the neoclassical ones, neither historical developments, nor political organization, nor sociological problems were considered to be relevant to the domain of the "universal"(?) theories of "scientific" economics. The promise of these theories, i.e. that the free markets would reach optimal levels by means of a divine contribution of an "invisible hand", unfortunately did not occur. Furthermore, growth problems had been continuing and threatening Global stability. Inevitably something had to be done.

In such an environment, the concept of "development economics" emerged as a way out. Since it is out of the question to abandon the Western scientific

---

23 Since we analyzed the causes of growth earlier in this book, it will be redundant to discuss it here again. However, very briefly, it will be convenient to emphasize again the importance of the technological inventions and qualifications of the labor using such technologies.

(?) growth theories on the grounds that they are "not functional", it was necessary to find new solutions to the "actual" problems of developing countries. The existence of a very serious and tangible reality was obvious: A serious gap in terms of the Global income levels between countries was a fact and this gap has been increasing. Who are the culprits? The traditional and "scientific" economic theories? Or the unskilled decision-makers in the developing countries who implement these so called Western "scientific" theories? Naturally, one would not expect any "scientific" (!) economist to admit that *"our universal scientific theories cannot, or are insufficient to, comprehend the real problems of the developing countries"*. Of course any realistic solution would require a comprehensive analysis of the historical, political, social, psychological, and cultural facts, which are ruled out by the Western "scientific" economic theories. As a result, a new concept and a sub-branch of economics had emerged; "development economics".

Elements of the Development Theory

Now, growth theories as well as development theories abound. There's nothing to do except move ahead. However we should point out the differences and similarities between the "scientific" (!) growth theories and the development theories so we don't get mixed up. Since we defined growth as an increase in the added value, so we have to define the notion of development clearly and then present the differences between these two concepts.

Development analysis is a process which embraces "change" just as the concept of growth does. Change, or rather transformation, is related to the ongoing institutional, political, social and cultural infrastructure of a country. The most important of them are factors such as individual rights, freedom, schooling rate, urbanization rate, population growth, environmental awareness, health expenditures and telephone or computers per capita. If we are to list these factors in relation to development, change is expected in the following areas.

Expectations from the institutional aspect:

1. Better and more transparent governance.
2. Implementing the 'appropriate' economic policies.
3. Increasing participation in the decision-making process.
4. Better supervision on the decision-makers and rulers.
5. Increasing safety mechanisms against the power of bureaucracy.
6. A fairer legal system.
7. Fairer regulations for tax collection.
8. A loan distribution system which ensures fairness and equal opportunity.

9. Fairer arrangements for revenue distribution.
10. Compliance with competition conditions in public procurement.
11. More respect for and measures to preserve ecological balance.
12. A fight against corruption.

Expectations from the individual aspect:

1. Enhancing individual rights and freedoms.
2. Increasing equality in educational opportunities (especially for the females).
3. Better quality of education.
4. Increasing measures against gender discrimination.
5. Higher levels of income.
6. Production and/or consumption of goods and a higher quality of service.
7. Decreasing the differences between urban-rural areas.
8. A fairer and better quality health system which protects low-income groups.
9. Putting an end to the outdated customs such as "honor killings", feuds".
10. Cleaner air.
11. Better living standards and environment.

No individual or institution of neither advanced nor developing country would probably object to the above expectations. All these expectations are the expectations of any rational individual in any developed or developing country. The difference between the advanced and developing countries is the fact that while advanced countries have made very considerable progress on all these items, developing countries have a long way to go and many progressive steps to take. In other words, the basic difference between advanced and developing countries is the quantitative and qualitative progress made by the advanced countries on the aforementioned items which places them ahead of the developing countries.

It is possible to derive the following conclusion from all that has been previously written: "the development process" continues in both developed and developing countries. This is simply because; development is a long-run and continuous process. By employing technological innovations which are the means we use to change and increase our control over the environment in which we live, not only does growth increase but also development proceeds. The reason behind this is that technological innovations lead to changes in production and consumption resulting in fundamental changes in the way we live. For example, widespread usage of the Internet not only influenced the information technology sphere but also changed exchange relations through e-commerce, communication through e-mail, the structure of individual-government relations through e-government in both advanced and developing countries. This process of change still continues.

Table: 5-1 Structural changes accompanied by the development process

	Institutional	Cultural	Economic	Political
Advanced Countries	Yes	Yes	Yes	Yes
Developing Countries	Yes	Yes	Yes	Yes

The conclusion to be drawn from all of this is; the growth and development process, e.g., transformation, has been in progress in both advanced and developing countries (Table: 5-1).

Development Economics

Developing countries have many complicated and serious problems which must be urgently remedied. If we are to list them, the most important are these:

1- Low level of income.
2- Unequal income distribution.
3- Corruption.
4- Inappropriate economic (micro-macro-foreign trade) policies.
5- Unequal educational opportunities (particularly for females).
6- Inadequate skill level of the labor-force.
7- Institutional irregularities.
8- Inappropriate science-technology policies.
9- Insufficient communication infrastructure.
10- Inadequate transportation infrastructure.
11- Bureaucratic oligarchy.
12- Rapid population rise.
13- Non-transparent governance.
14- Problems concerning supervision.
15- Decision mechanisms not open to public participation.
16- Customs specific to underdevelopment.
17- Restrictions on individual rights and freedom.
18- Inadequacy of democratic culture.
19- Unplanned urbanization leading to slum areas.
20- Developmental differences between urban and rural areas.
21- Low per capita efficiency in agricultural production.
22- The relative size of the agricultural population.
23- Inefficient and inadequate health services.

24- Air and environmental pollution.
25- A deteriorating ecological balance.
26- Gender discrimination.
27- Class privileges (chieftainship, tribes, caste systems, etc.).
28- Restrictions on cultural rights and freedom.

Once they are scrutinized, it will be evident that the growth theories of the neoclassical doctrine cannot be a remedy to any of the above underdevelopment problems. As we recall, the neoclassical doctrine inherently ignores such problems.

In such an environment, the introduction of "development economics" seemed a reasonable way out of the prevailing difficulties. In other words, in an environment where 'traditional' growth theories do not seem to address the actual issues in developing countries, it was inevitable that an approach which aimed to introduce realistic remedies to the existing problems should be adopted. *Development economics* tries to *discuss the "real" problems* of developing countries *in a more "realistic" manner.*

Concisely, we can define the "development economics" as a sub-branch of economics dealing with the transformation of *"qualitative socio-economic aspects"* accompanied by the growth process.

Chapter-11
Epilogue & a Suggestion

The "neoclassical" theories sprung up around the 1870s and have flourished since then. Initially, as they appeared to use the advanced mathematical models and employed "universally binding laws" like physics and astronomy, these theories were believed to be a pure science. One of the most significant findings of this study on growth is that they were pseudo "academic" studies which analyzed the unrealistic and fictitious relationships of an "imaginary" economic world. None of the other social sciences, claim, either to employ advanced models of calculus, or to be a positive natural science as does economics. It does not mean that these neoclassical theories, constructed from utopian hypotheses, using the logic of Newtonian mechanics in the quest for an imaginary "equilibrium", are totally useless. Once these theories are recognized and acknowledged as containing "normative", but not "positive" values, and then it may bear fruit and flourish. A totally different form of economics may emerge once that is seen in terms of "it should be" rather than "it-is".

In addition to criticism of many new and old theories in previous sections, the process of economic growth was considered under two headings, i.e., as short and long-run growth. In this book an attempt has been made to examine a more realistic method of analysis, and an alternative growth model. According to our findings short term growth is possible with some efficiency (micro-productivity) increasing precautions; however long-run growth is more important. Technological innovations appear to be the basis for long-run growth, and the source of technological innovations appears as *qualified laborers* or more precisely "*the creative minded laborer*".

When technology is evaluated from Global perspective, the countries, which have realized considerable technological innovations, achieve growth both slowly and quickly. So even though some problems are experienced in terms of unemployment, growth still continues. And, although developing countries have a big potential for growth and several countries have to some extent succeeded in closing the development gap between themselves and the developed countries, the income gap between most of the developing countries and the developed countries is continually growing. In other words, Global inequality grows constantly, and never appears to decrease. As Stiglitz highlighted, the gap between the rich and the poor is growing increasingly and even the number of those living in absolute poverty (people living on less than one dollar a day) does not decrease, but increases (Stiglitz;2002;46). The fact that a few countries increase their production and their relative welfare share in the existing Global order does not mean that the system functions well. On the contrary, it merely indicates that the global order does not work at all well.

At the root of the problem, i.e., that some countries are richer and some are poorer and/or they grow at different rates, lay the fact that they display differences in the qualification level of their labor forces as well as differences in the technological development of their companies or the country as a whole. In order for these differences to decrease and be erased in Global terms, first and foremost, it is necessary to have access to a labor force with appropriate qualifications that can use the given technologies efficiently in terms of production. Additionally, a restructuring that will ensure the further development of the technologies that are currently used is very important. However the firms located in the developing countries must firstly be a good "technology user" themselves in regard to their priorities and the Global realities before being a "technology producer". A properly qualified labor force is the indispensable prerequisite for such a long run growth.

It is a fact that developing countries have very significant problems regarding the qualifications of their labor force. The labor force of many countries cannot even use the existing technologies efficiently. When Global production relationships and "imperfections in the technology markets[24]" are combined with the poor use of existing technology, we observe that there are many significant problems related to Global growth.

Developed Countries & Long-run Growth

Growth needs to be evaluated differently from the point of developed and developing countries. The suggestions that are made when the developed economies and their relationships are in question are comparatively simpler: As long-run growth has its source in technological innovations and technological innovations derive its source from the creative intellectual labor, technological innovations must be encouraged; more importantly, the qualification level of the labor force must be improved. But most importantly, an education and production system that further features and awards "*creative*" intellectual activities must be promoted.

---

24 Technology market imperfection implies that the transfer of technology by the market is not efficient. That is, there exists other conceivable outcomes where the recipient country may be made better-off.

Developed countries already have a relatively good education system and a structure encouraging technological progress. They seem to have the necessary knowledge level, the organizational structure and the awareness to further advance their technology. Of course, there is not a directly proportional relationship between the improvement of the labor force's education level, R&D incentives, and new technological inventions. A guarantee cannot be given that better education and increased incentives will lead to the same level of technological innovation for everyone. However it is well known that there is a close and direct relationship between more *"knowledgeable creative minds"*, more R&D and more technological innovation.

LDCs & Long-run Growth

The differences in Global development levels, of course, are not something that emerged in a short period of time, say within several weeks, months, or even years. To understand the present day, the past 200 – 250 years must be thoroughly analyzed. What can or should developing countries do from now on?

It is not only the economic factors, but also the political, institutional and cultural factors which should be taken into account. Political, institutional and cultural structure exerts a great pressure on economic growth.

From an economic perspective; long run growth is a prerequisite for the realization of the long-run "development" process. To achieve long-run growth;

1. there must be technological innovations; and
2. qualified human resources that can use these technological innovations effectively.

Producing New Technologies

Developing countries, as distinct from developed countries, do not need to invent all the technologies that they require for growth from scratch. Because developing countries can tap into an enormous productive knowledge (technology) repository which has been accumulated by the developed countries over the previous centuries which they can also use for a long period of time, at least in theory. Developing countries have the potential to obtain a lot of the technologies which they require through technology transfer from the developed countries. However, as is well known, imperfections in the technology markets and that the Global production relations are being shaped more and more by "Giant Global Firms", tend to cause significant obstacles to the technology transfer process in real terms. In order to make technology transfer happen successfully, new Global technology policies which pay sufficient attention to the interests of developing countries are imperative. The existing order and the continuing globalization process do not appear to permit such a process to be successful.

More Efficient Use of Technologies

Using existing technologies and new technologies effectively is so important from a Global competition and growth point of view. To state the obvious, developing countries have serious shortages in terms of both the level of qualifications and quantity of the qualified labor force. On the other hand, the "brain drain" causes the qualified labor force shortage to intensify. While the quality of the labor force has to be urgently improved, accurate measurements have to be taken globally in order to prevent a "permanent" brain drain which would favor the developed countries, and at the same time a *"reverse brain drain"* in favor of the developing countries has to be encouraged.

To train the available labor force in the developing countries to use the existing technologies and the transferred advanced technologies efficiently is a problem that has to be dealt with, primarily, by their rulers. To this end, attention must be paid to education, especially in the fields of training in technical issues, and the necessary resources must be available for this. International organizations and establishments or those globally operating enterprises located in developed countries may make significant contributions in this endeavor. However even if solving the education-training problems of the labor force seems easier than the elimination of the imperfections in the technology markets, a huge educational investment has to be made. Even with a huge investment it may take a long time before any significant results are realized.

Efficient Use of Political, Institutional & Cultural Framework

Political, institutional and cultural structures constitute critical elements in the economic performance of a country. For example, political and economic decisions taken by the rulers of a country may pave the way for or hinder the entrepreneurs of that country. For instance, an important part of the financial resources of Turkey were wasted due to financing malpractice instead of being used to increase production due to the improvidence of its rulers in the 1990s. The malpractice in the banking sector alone was estimated to be 40 – 50 billion US Dollars.

Not only political or institutional corruption but also the cultural structure has an impact on the emergence of malpractice. Nevertheless, with regard to the global economic structure and its relationships, and the conditions of our era, the problem cannot be overcome simply by changing the institutional and cultural infrastructure of a country. Global economic relations must be reviewed completely and there is an urgent need of a "Global institutional re-structuring" in order to increase the technology transfer to the firms in developing countries.

A Suggestion to Boost Global Cempetition

There are three unquestionable economic realities:

1. Competition is beneficial.
2. For (Global) competition, contemporary "technology" is required.
3. Without a qualified labor force, competitive technologies cannot be used effectively.

One of the most useful things introduced by the "utopian" Neoclassical doctrine, which is still found in present-day text-books, is that the markets will be at "optimum" level when the number of producers is as many as the atoms, no one controls the market, the preferences of the consumers reign supreme, knowledge is freely available. In this perfect competitive environment, no producer is capable of getting monopolistic or oligopolistic power or control, and the consumers get the most benefit.

However the fact that "perfect" competitive markets are or have never actually been present in "real" economies; the Global production relations of our era tend to decrease the competition. An important portion of global production is confined to firms having global operations, and these are mostly rooted in the developed countries. The amount of production and distribution under their control is increasing. In other words, oligopolistic or monopolistic control tends to grow stronger in the areas of Global production and distribution, whereas Global competition tends to decrease accordingly. For example, in the 1980s, the number of automobile manufacturers that had Global competitive power was limited. Within the past 30 years, many automobile firms were acquired by the rival automotive companies located mostly in the developed countries. As a result the number of competing independent automobile manufacturers decreased constantly which led to decreased Global competition, while oligopolistic power increased.

The things experienced in the automobile industry are also observed in highly dynamic sectors which employ the most advanced technologies. Either globally powerful companies originating in developed countries take over their rivals one by one. Some developing country companies do not even attempt to compete on account of the differences in technological development levels. For example, firms in the developing countries generally have neither the power to compete nor the capacity to produce contemporary technologies in fields such as biochemistry, gene technology, communication technologies, or the medical industry, where the most advanced technologies are required. In such a case, the best thing that the firms in developing countries can do is to cooperate with companies which have advanced technologies and take part in Global production by manufacturing the components requiring a "cheap labor-cost". This process is often supported by generous financial incentives provided by the respective governments to their foreign investors. However there is no guarantee of permanency in this kind of partnership. When the government incentives diminish or are eliminated, or other countries present more attractive incentives, especially when the charm of *"cheap labor"* disappears, Global production will be restructured in accordance with the global interests of the Giant Firms.

While the firms in the developed countries are granted many privileges within the globalization process, technology transfer to developing countries is often very limited; therefore the environment that allows the development of the firms to compete independently and globally in the developing countries is not created. Consequently the economic dependency of developing country firms on the developed country firms increases with globalization, while Global competition gradually decreases.

If competition, is in fact "perfect" as taught by neoclassical economics and is the most beneficial one to consumers; why, in order to increase Global trade aren't measures taken by the developed countries or Global institutions to increase global competition? Why are the Global oligopolistic and monopolistic trends overlooked or ignored?

Why is it not possible to do something to change this situation?

In regard to the last question, a possible resolution is discussed below which could be useful in global terms and may boost global competition.

Firms that "Only Produce Technology"

As it is known, the technological innovations which are used or required by private companies are generally acquired by:

1. R&D studies from the producer firm
2. As a result of "common" R&D studies conducted with other producer firms universities and/or public institutions.
3. Placing an order with private or public R&D organizations.
4. Purchasing the patent of others' inventions.
5. Leasing "use" rights by means of a license.

The technology producer firm holds the ownership of the technology, i.e. the production-related knowledge and also has the monopoly power until its rivals develop similar products. In the meantime, it can control the markets however it wishes and determine the price and/or the supplied quantity. The profit rate to be obtained by this firm by means of the monopoly power provided by right of ownership is expected to be higher than the average rate in the market.

That is because there is no competition.

Yet, competition in terms of the economic efficiency and consumer welfare is more useful than a monopoly; at least that's what the Western born economic doctrines say; especially those that stress the importance of competition.

Is it possible to increase Global competition, while protecting the right of ownership of technological innovations through the patent laws?

The answer is "Yes". And in such a way not only the "technology producing firm" that holds the patent of the technology, but all the other firms and countries globally may benefit from it. It is a very simple suggestion:

If the technology producing firms and the technology using firms, operate as profit units that are totally independent of each other, both global competition will increase and most of the problems in the technology markets will disappear, ensuring the rapid improvement of global welfare, hence reduce the gap in welfare levels among countries.

Here is an example to illustrate the situation better: Let's assume that R&D firm "X" invents and develops a solar energy engine which functions in the same way as the hybrid engines. The R&D firm X is entitled to grant a voluntary permit (lease) to any firm who wishes to produce the solar energy engine in their own production facilities. The more companies that R&D firm X grants the license of its own invention to, the more license-fee it will earn. As a result, as each firm will have access to the new technology and global competition will increase.

The increase in Global competition is not expected to be as easy and smooth as it is in the neoclassical "free market" models. In other words, not every firm that wants to will be able to manufacture Firm X's engine. Among tens, even hundreds of thousands of producer firms that are scattered around the world, maybe very few of them have the adequate infrastructure and labor force to benefit efficiently from new technology. Nevertheless, the competition will be more severe than the current competition and the market will be always "potentially" ready for the entry of new producers. As the different levels in terms of qualifications of the labor force even out between the countries and the firms in developing countries increase their competitive power and gain experience, Global competition will rise and the consumers will benefit greatly.

This suggested solution may sound "utopian" or "highly optimistic" to some. In fact, even if it is not the best fit; operating systems produced by Microsoft for PC's are a good example. Microsoft sells the Windows operating systems they produce to any applicant firm producing a computer. As a result serious Global competition started between computer producing firms and consumers made the most of it.

The globally operating firms who are getting the maximum benefit from the existing system will obviously object to such an arrangement. Why would they want to change "ex nihilo" *a system which favors their interests?*

On the other hand, even if such an environment as proposed emerges, many firms, especially those in the developing countries will not be able to benefit from the new system immediately. The main reason for this is that the technological and organizational infrastructure as well as the quality of the labor force of each firm is not appropriate to benefit from it. For example, even if Airbus is ready to license out all its innovations to every applicant company, the number of the companies that have the capacity to produce the aircraft is limited.

The biggest 100 firms in the world have their origins in developed countries and their annual turnovers are bigger than many countries' GNP. In order for Global competition to be fairer and more efficient, financial and organizational differences between firms also need to diminish. This will of course take a great deal of time.

Because the economies of countries grew at different rates in the past, the natural outcome was a difference in their income. Unfortunately this gap does not tend to diminish, but instead, as pointed out before, it grows. However; welfare differences between countries is not simply "fate", nor do some countries "deserve" to be poor. These differences were created by "man" and may be eliminated by "man", if he takes the necessary precautions.

In short, in order to increase Global competition:

1. *Global production and distribution relationships*; and
2. *Global technology transfer policies and technology markets*

SHOULD BE RESTRUCTURED but this time with positive discrimination in favor of the developing countries.

Suggestions for "Fairer" Income Distribution & Increased Democracy at Work

All governments around the world, with the exception of one or two, frequently grant various and sometimes extremely generous, incentives to private sector investment in order to promote growth, increase employment, exports and regional development. These incentives can be in the form of loans at a low rate of interest, tax-holidays, subsidies to employ, etc. Normally these incentives make the capital-owner richer than he/she was before.

Our suggestion is related to these financial incentives. If and when a firm is granted financial benefits of this kind, at least half of the incentives should be given to the employees to buy "shares" in the firm. For example, if the firm X gets a 100 million Euros credit at low interest rate, 50 million Euros should go to the employees to buy shares and become partners.

Certainly not all of the employees may be anxious to buy shares which would make them a partner of a potential loss as well. But, at least, there will be a choice for them which could help to increase the productivity, democracy at the plant, fairer distribution of the income between partners of production. As long as the firm continuous to make profits, the income of the employees with shares would increase. Meanwhile, the loyalty to the firm would increase and the employees would work more efficiently. In troubled times, the employees with shares would be more inclined to cooperate for the survival of the firm.

Is "Unlimited Growth" Desirable?

According to one of the significant assertions, a creative human mind has an unlimited capacity to create new ideas which can be transformed into products with an exchange value. This creative capacity implies continuous introduction of "new" technologies which secure long-run growth.

In addition, it is commonly acknowledged that the firms introducing "new" products through new technologies can influence, if not manipulate, the human mind to buy more and more products through various means, mainly by advertising. For example, assume that a family with two kids has two TV-sets at home and they easily meet the requirements of the household members, although there might sometimes be small problems on the choice of program. If the father and the son want to watch the football game while the mother and the daughter prefer to watch a TV-movie, there would not be any problem. We ignore the social and physiological impacts of a family-split in the same house.

Given the potential purchasing power supported by financial incentives, the family of the four can be influenced (manipulated) to buy a separate TV-set for the young girl of the house. If the young girl gets it, why doesn't the young boy get one, too? After all, most of us, thanks god not all of us, are already prepared to believe that more consumer goods implies more happiness, more social status, and more satisfaction. If the possessions are more up to date than the possessions of the people around us, the greater the satisfaction and happiness is believed to be.

The same feelings and attitudes apply for many other products such as smartphones, tablets, automobiles, etc. The more up to date the product, the more happiness (!); at least many "modern consumers" think so. It doesn't matter much whether the consumer really needs them, nor is the issue whether the one we already possess meets all our requirements or not. Happiness and social status is correlated with the products we possess, unfortunately.

This passionate demand for more and newer products is an absolutely fine consumer behavior appreciated by the producers and sellers. But what about the social, psychological and environmental aspects of this unrestrained hunger for consumption? Everybody is, more or less, familiar with the destructive environmental aspects in terms of worsening quality of air, water, food, etc. As "rational" consumers, are we going to continue living as today until we destroy the final tree, river or animal?

Let's consider the human relations aspect. Are we aware of the negative effects of our voracity on other human beings? Or how much do we care about it? We frequently observe that the car-owners spend more time on their possessions than the time they spend with their children or friends. The time we "kill" in front of the TV or computer is invaluable but are we aware of it? We are persuaded to think that "more" and more up-to-date possessions provide more happiness. But do we know what is worth to spend time or do something together with the loved ones in terms of happiness? Moreover, aren't we responsible for the people starving or living in poverty? How much do we care about them when we go crazy about having more and more possessions? How can we sleep in peace when all these global problems exist?

In various parts of this book, we claimed that there is no upper limit in the "creative capacity of human mind (labor)". Thus, there is no potential upper limit to long-run growth, as long as "new" technologies are developed.

But, is it not high time to use the "creative capabilities" of human brain to help to solve the global problems of mankind?

Bibliography

Adelman, I. (1972): Ekonomik Büyüme ve Kalkınma Teorileri Bursa İTİA, Yayın no: 3

Acar, G.T. (2008): İktisadı Değiştirmek İletişim Yayınları, İstanbul.

Acemoglu, D- Johnson, S.–Robinson, J. (2004): "*Institutions as the Fundamental Cause of Long- run Growth*" *NBER* Working Paper-10481

Acemoglu, D. ve Robinson, J.A. (2013): Ulusların Düşüşü Doğan Kitap, İstanbul.

Acemoglu, D. (2008): Introduction to Modern Economic Growth Preliminary Electronic Ed., Princeton Uni. Pres.

Aghion, P. – Howitt, P. (1992): "*A Model of Growth Through Creativ Destruction*" *Econometrica*; Vol.60; No:2

– (2009): The Economics of Growth MIT Press.

Ashall, F. (2006): Olağanüstü Buluşlar TÜBİTAK popüler Bilim Kitapları: 208

Akat, A.S. (2004): "*Alternatif Büyüme Stratejisi*" http://akat.bilgi.edu.tr/pdf/alternatif-2.pdf; 2008-12-13

Al-Suwailem, S. (2008): Islamic Economics in a Complex World Islamic Development Bank.

Aren, S. (1998): Istihdam Para ve Iktisadi Politika Savaş Yayınevi, Ankara.

Baldwin, R.E. (1982): "*Gottfried Haberler's Contributions to Int. Trade Theory and Policy*" The Quarterly Journal of Ec.", Vol.97; No.1

Barro, R. J. (1990): "*Government Spending in a Simple Model of Endogenous Growth*". Journal of Political Economy; Vol.98.

– (1995): "*Inflation and Economic Growth*" NBER Working Paper No:5326

– (1999): Determinants of Economic Growth The MIT Pres.

– (2000): "*Inequality and Growth in a Panel of Countries*" Journal of Economic Growth; 5: 5-32

Barro, R.J. – Sala-i-Martin, X. (1991): "*Convergence Across States and Regions*" Brookings Papers on Economic Activity; Vol.1991; No.1; pp.107-182

– (2004): Economic Growth, 2. Ed. The MIT Pres.

Başkaya, F. (1994): Kalkınma İktisadının Yükselişi ve Düşüşü İmge Yayınevi.

Baumol, W.J. – McLennan, K. (Eds.) (1985): Productivity Growth And US Competitiveness Oxford Uni. Press, N.Y.

Baumol, W.J. (2008): "*Entrepreneurship and Innovation: The (Micro) Theory of Price and Profit*" www.aeaweb.org/annual_mtg_papers/2008/ 2008_345.pdf

Becker, G.S. (1975): Human Capital. National Bureau of Economic Research, N.Y. Mc Graw-Hill Book Co., New York.

Bernard, A.B. – Jones, J.I. (1996): Productivity Across Industries & Countries *The Review of Ec. & Statistics*, Vol:78, No:1

Bhagwati, J. (1958): The Immiserizing Growth: A Geometrical Note" *The Review of Econ.Studies*, Vol.25; No.3, June

Binder, A.S. (2000): "*Keynesyen İktisadın Düşüşü ve Yükselişi*" Ö.Demir Ed. İçinde; Devlet, Rekabet Mülkiyet ve İktisat. Değişim Yay.

Blaug, M. (1963): "*A Survey of the Theory of Process-Innovations*" *Economica*, Vol.30; No: 117; 13-32

– (1980): The Methodology Of Economics. Cambridge Uni. Press, Cambridge.

– (1990): The History of Economic Thought. Edward Elgar Pub.. Ltd., Hants.

Bohman, R.S. (1990): "*Smith, Mill & Marshall On Human Capital Formation*"; *History of Political Economy*; Vol. 22:2; Brookings Papers on Ec. Activity.

Bolio, E. et. al. (2014): "*A tale of two Mexicos: Growth and prosperity in a two-speed economy*" http://www.mckinsey.com/Insights/Americas/A_tale_of_ two_Mexicos?cid=other-eml-alt- mgi-mck-oth-1403, 2014-04-27, McKinsey Global Institute.

Brecher, R.A. – Choudhri, E.U. (1982): "*Immiserizing Investment from Abroad: The Singer-Prebisch Thesis Reconsidered*". *The Quarterly Journal of Econ.*, Vol.97; No.1

Bruno, M. – Easterly, W. (1995): "*Inflation Crises and Long-run Growth*" NBER Working Paper No:5209

Buğra, A. (1989): İktisatçılar ve İnsanlar Remzi Kitabevi, İstanbul.

Bulutay, T. (1972): Iktisadi Büyüme Modelleri Üzerine Açıklamalar ve Eleştirmeler. Ankara Üni. SBF No:341

Bülbül, Y. (2008): Teknonomi Kitabevi, Istanbul.

Caballe, J. – Santos, M.S. (1993): "*On Endogenous Growth With Physical & Human Capital*" *The Journal of Pol. Ec.*, Vol.101; No:6, Dec.

Capolupo, R. (2008): *"The New Growth Theories and Their Empirics after Twenty Years" Economic Discussion Paper* 2008-27 www.economics-ejournal.org/ economics/discussionpapers/2008-27

Cetindamar, D. - Phaal, R - Robert, D. (2013): Teknoloji Yönetimi Efil Yayınevi, Ankara.

Chandrasekhar, C.P - Ghosh, J. (2013): *"Do Wage Shares Have to Fall with Globalisation?"* www.networkideas.org/news/jul2013/Globalisation.pdf, 2013-08-28

Chang, H.J. (2008): *"Kicking Away the Ladder"* www.btinternet.com/~pae_news/ review/issue15.htm; 2008-11-09.

Chang, H-J. - Grabel, I. (2005): Kalkınma Yeniden İmge Kitabevi, İstanbul. Çev. E. Özçelik

Cervantes, M. - Guellac, D. (2002): The Brain Drain: Old myths, new realities www.oecdobserver.org/news/fullstory.php/aid/673

Cohen, A.J. - G.C. Harcourt (2003): *"Whatever Happened to the Cambridge Capital Theory Controversies?" Journal of Economic Perspectives.* Vol. 17 No: 1

Collins, S.M. - Bosworth, B.P. (1996): *"Economic Growth in East Asia". Brookings Papers on Ec.Activity;* Vol.1996

Coppel, J. - Dumont, J.C. - Visco, I. (2001): *"Trends in Immigration and Economic Consequences"* OECD-Working Paper No: 284: www.oecd.org/dataoecd/29/30/1891411.pdf, 2005-07-13

Cornforth, M. (1975): Bilgi Teorisi Ma-ya Yayınlari, Istanbul.

Clark, J.B. (1894): *"The Genesis of Capital". Publications of AEA,* Vol.9; No.1; 64-68

Crawford, R. (1991): In the Era of Human Capital. Harper Business, New York

Deliktaş, E. (2001): Malthusgil Yaklaşımdan Modern Ek. Büyümeye *Ege Akademik Bakış,* Cilt:1, Sayı:1

De Long, J.B. (1992): *"Productivity Growth & Machinery Investment: A Long Run Look, 1870-1980" The Journal of Ec. History;* Vol.52; No.2; pp.307-324

– (1997): *"What Do We Really Know About Ec.Growth?"* www.j-bradford-delong. net/Econ_Articles/hoover/growth_delong_hoover.html

De Tarde, G. (2004): Ekonomik Psikoloji, 1. Cilt Öteki Yayınevi, Ankara. Çev. Özcan Doğan.

Diamond, J. (2004): Tüfek, Mikrop ve Çelik - İnsan Topluluklarının Yazgıları (Guns, Germs and Steel) TÜBİTAK Popüler Bilim Kitapları 174

Dornbush, R. - Fischer, S. (1998): Makroekonomi Akademi Yay. Hizm. Ankara

Drucker, P.F. (1981): Toward The Next Economics Harper & Row Publ., New York.

– (1993): Yeni Gerçekler İş Bankası Kültür Yayınları No: 315

– (1995): Gelecek İçin Yönetim. (Managing For Future) İş Bankası Kültür Yayınları No: 327

Duraiappah, A.K. (2000): Sustainable Development & Poverty Alleviation Int. Inst. For Sustainable Development

Easterly, W. (1998): "*The Quest for Growth*" worldbank.org/research/growth/notes1.html

– (2005): "*What Did Structural Adjustment Adjust?*" *Journal of Development Ec.*, 76,

Easterly, W. & et.al (1994): "*Policy, Technology Adoption & Growth*" *NBER Working Paper*; No: 4681, March-1994

Eichengreen, B. - Park, D.- Shin, K. (2013): "*Growth Slowdowns Redux*" NBER Working Paper No. 18673

Eichner, A.S. (ed.) (1979): A Guide to Post-Keynesian Economics M.E. Sharp Inc. New York.

Ellsworth, P.T. (1940): "*A Comparison of International TradeTheories*" *AER*; Vol.30; No.2

Eren, E. - Sarfati, M. (Ed.) (2011): İktisatta Yeni Yaklaşımlar İletişim, Istanbul.

Erkan, H. (1994): Bilgi Toplumu Ve Ekonomik Gelişme. İş Bankası Kültür Yayınları No: 326

Fagerberg, J. (1994): "*Technology and International Differences in Growth Rates*" *Journal of Economic Literature*; Vol.32; No.3

Fisher, S. (1993): "*The Role of Macroeconomic Factors in Growth*" *NBER Working Paper*: No.4565

Frankel, J.A - Romer, D. (1999): "*Does Trade Cause Growth?*" *AER*; Vol.89; No.3

Frankenhoff, C.A. (1962): "*The Prebisch Thesis: A Theory of Industrialism for Latin America*". *Journ.of Inter-American Studies*, Vol.4; No.2

Freeman, C. - Soete, L. (2003): Yenilik İktisadı (Çev. E. Türkcan) TUBITAK, Ankara.

Goodwin, N. - Nelson - J.A. Haris, J. (2008): Macroeconomics in Context M.H. Sharp

Goodwin, N. & et. al. (2008): Microeconomics in Context M.H. Sharp

Gorz, A. (2007): İktisadi Aklın Eleştirisi Ayrıntı yayınları, Istanbul. Çev. I. Ergüden

Griliches, Z. (1997): "Education, Human Capital & Growth" Journal of Labor Economcs, Vol:15, No:1, Jan.

Grandville, O. (Ed.) (2012): Economic Growth and Development EBSCO Publishing

Grossman, G.M – Helpman, E. (1990): "Comparative Advantage and Long-run Growth", AER; Vol.80; No.4.

– (1991): Innovation and Growth in the Global Economy MIT-Press, Cambridge.

Gürak, H. (1990): Transfer Of Technology Unpublished thesis; Uni. Of Lund, Sweden.

– (1993): An Alternative Price Theory. Unpublished Docent Thesis,

– (1999)-a: "Ülkelerin Refahının Kaynagı Nedir?" Banka ve Ekonomik Yorumlar; Ocak-1999

– (1999)-b: "On Productivity Growth" YK-Economic Review; Dec.; Istanbul.

– (2000)-a: "Economic Growth and Productive Knowledge" YK-Economic Review; June; Istanbul.

– (2000)-b: "Verimlilik Artışları" Verimlilik Dergisi; Eylül-Ekim; Ankara.

– (2003): "Hidden Costs of Technology Transfer" YK-Economic Review; June; Istanbul.

– (2004): "On Value and Price" YK-Economic Review; June; Istanbul.

– (2007): "Kutsal İdeolojinin Eleştirisi" veya "Bilimsel (!) İktisadın Sefaleti". www. hasmendi.net, 2013-05-21

– (2012): Heterodox Economics Peter Lang, Frankfurt.

– (2013): Heterodox Economics-2 Peter Lang, Frankfurt.

Güvel, E.A. (2011): Ekonomik Büyüme Kuramları Karahan Kitabevi, Adana.

Han, E. – Kaya, A.A. (2006): İktisadi Kalkınma ve Büyüme Anadolu Üni. Yayını, No: 1575

Harrod, R.F. (1937): "Mr. Keynes and Traditional Theory". Econometrica, Vol.5; No.1, Jan. 74-86

– (1973): Economic Dynamics. Macmillan Press, London.

Hausman, D.M. (1981): Capital, Profits and Prices. Columbia Uni. Press, New York.

– (1988): "*Ceteris paribus Clauses and Causality in Economics*" Proceedings of the Biennial Meeting of the Philosophy of Science Assoc. Uni. Of Chicago Press.

Helleiner, G.K. (1981): Intra-Firm Trade & The Developing Countries St. Martin's Press, New York.

Hicks, J.R. (1965): Capital And Growth. Oxford Uni. Press, London.

– (1966): "*Growth and Anti-Growth*" Oxford Economic Papers, Vol.18; No.3; 257-269

– (1970): "*A Neo-Austrian Growth Theory*" The Economic Journal, Vol.80; No.318; June.

– (1979): Causality in Economics. Basic Blackwell, Oxford.

– (1981): "*The Mainspring of Economic Growth*" AER, Vol.71; No.6; 23-30

– (1983): Classics and Moderns. Basil Blackwell Publ., Oxford.

Hikino, T. (2005): Business Organizations & Modern Ec. Growth. Yasar University, Int. Conference.

Howitt, P. (2000): "*Endogenous Growth and Cross Country Income Differences*"; The AER; Vol.90; No.4; 829-846

Hoover, G.E. (2008): "*The Present State of Economic Science*" Econ Journal Watch; Vol.5; No.3; 2008-10-12 www.econjournalwatch.org/pdf/HooverCharacterIssuesSeptember2008.pdf

Hultman, C.W. (1967): "*Exports and Economic Growth*" Land Economics, Vol:43, No:2, May-1967

Hume, D. (1986): Insan Zihni: (Essays concerning the human understanding) Milli Egitim Basimevi, Istanbul.

İnal, V. (2013): Büyüme Teorisinin Gelişimi ve Türkiye'nin Büyüme Sorunları Efil Yayınevi, Ankara.

İnalcık, H. (2000): Osmanlı İmparatorluğu'nun Ekonomik ve Sosyal Tarihi, I. Cilt Eren Yayıncılık, İstanbul.

İnsel, A. (2003): İktisat İdeolojisinin Eleştirisi Birikim Yayınları, İstanbul.

İrmiş, A. (2008): Küreselleşme Sürecinde Yeni Gelişen Piyasalar ve KOBİ'ler. www.hasmendi.net; 2008-01-14

Johnson, L. (1977): *"Keynesian Dynamics and Growth" Jour.of Money, Credit & Banking,* Vol.9; No.2;

Jones, C.I. (1997): *"On the Evolution of the World Income Distribution" The Journal of Ec. Perspectives;* Vol.11; No.3.

Kaldor, N. (1957): *"A Model of Economic Growth" The Economic Journal;* Vol.67; No.268.

− (1960): Essays On Value And Distribution. G.Duckworth & Co. Ltd., London.

− (1964): *"International Trade and Economic Development" The Journal of Modern African Studies;* Vol.2; No.4

− (1985): Economics Without Equilibrium. University College Cardiff Press, Cardiff.

− (1986): *"Limits on Growth" Oxford Ec. Papers,* Vol.38; No.2; 187-198

− (1989): Further Essays On Economic Theory & Policy Duckworth, London.

Kaldor, N. − Mirrlees, J.A. (1962): "A New Model of Economic Growth", *The Review of Economic Studies;* Vol.29; No.3.

Kalecki, M. (1987): Selected Essays on the Dynamics of Capitalist Economy 1933-1970 Cambridge Uni. Press.

Karluk, S.R. (2005): Türkiye Ekonomisi'nde Yapısal Dönüşüm Beta, İstanbul.

Keesing, D.B. (1965): *"Labor Skills and International Trade" The Review of Ec. & Statistics,* Vol.47; No.3

− (1966): *"Labor Skills and Comparative Advantage" AER,* Vol. 56; No. 1/2; 249-258

Keynes, J.M. (1971): The Economic Consequences of the Peace Macmillan, Cambridge Uni. Press

− (1973): The General Theory of Employment, Interest and Money. Cambridge University Press.

− (1978): *"Some Economic Consequences of a Declining Population" Population and Development Review* Vol.4; No.3; 517-523

Knight, F.H. (1964): Risk, Uncertainty and Profit. Sentry Press, New York.

Korkmaz, A. & et.al (2013): İnsani Ücret İLKE Araştırma Raporları:3, İstanbul.

Köse, A.H. (1992): Büyüme ve Verimlilik MPM Yay. 471, Ankara.

Krugman, P.R. (1980): *"Scale Economies, Product Differentiation, and the Pattern of Trade", AER,* Vol.20; No.5; 950-959.

– (1983): *"New Theories of Trade Among Industrial Countries"*, The AER, Vol.73; No.2.

– (Ed.) (1986): Strategic Trade policy and the New Inter national Economics. The MIT Press, Cambridge

– (1996): Rethinking International Trade The MIT Press.2008-11-10.

– (2008): *"Fire-Sale FDI"* http://web.mit.edu/krugman/www/FIRESALE.htm

Krugman, P.R. – Obstfeld, M. (2008): International Economics: Theory & Policy Pearson Education.

Kuhn, T.S. (1982): Bilimsel Devrimin Yapisi. (The structure of scientific revolutions) Alan Yayincilik, Istanbul.

Laczko, F. – Gozdziak, E. (Eds) (2005): Data and Research on Human Trafficking: A global survey, Int. Org. for Migration, Geneva.

Laitner, J. (1993): *"Long-run Growth and Human Capital "* The Canadian Jour. Of Ec.,Vol:26, No:4

Landau, R. – Taylor, T. – Wright, G. (1996): The Mosaic of Economic Growth Stanford Uni. Press.

Landsburg, S.E. – L.J. Feinstone (1997): Macroeconomics McGraw Hill Co., New York.

Leigh, A.H. (1974): *"Frank H. Knight as Economic Theorist"* The Journal of Political Economy; Vol.82; No.3

Lenihan, J. (2005): Bilim İş Başında TÜBİTAK Popüler Bilim Kitapları: 113

Lester, R.A. (1946): *"Shortcomings of Marginal Analysis for Wage-Employment Problems"* AER, March 1946.

Leontief, W. (1956): *"Factor Proportions and the Structure of American Trade"* The Review of Economics and Statistics; Vol. 38; No.4; Nov.; 386-407

Liagouras, G. (2008): *"American Institutionalism"* http://eaepe2008.eco.uniroma3.it/index.php/eaepe/eaepe2008/paper/viewFile/225/98 2008-11-18.

Lipsey, R.G. et.al. (1990): İktisat Bilim Teknik Yayınevi, İstanbul.

Lucas, R. (1988): *"On The Mechanics Of Economic Development"*. Jour.of Monetary Economics; July; 1988.

Machlup, F. (1935): *"Prof. Knight and the 'Period of Production'"* The Journal of Pol.Ec., Vol.43; No.5; 577-624

Mankiw, G. N. (1995): The Growth of Nations Brookings Papers on Economic Activity The Brookings Institution, Sept. 1995

– (2003): Macroeconomics Worth Publ.

– (2004): Principles of Economics Harvard Uni.

Mankiw, G.N. – Romer, D. (Eds) (1991): New Keynesian Economics MIT Pres, Vol. I & II.

Mankiw, G.N. – Romer, D. – Weil, D.N. (1992): "*A Contribution to the Empirics of Economic Growth*" The Quarterly Journal of Economics; Vol.107; No:2; p. 407-437.

Marshall, A. (1961): Principle of Economics, Vol. 1 & 2 Macmillan And Co., London.

Marx, K. (1976): Capital, Vol. I Penguin Books.

– (1977): Capital, Vol. II Lawrance & Wishart, London.

– (1981): Capital. Vol. III Penguin Books.

Massey, R. (2006): Microeconomics in Context Preliminary Electronic Ed., Tufts Uni., GDAE Inst.

Mathur, V.K. (1991): "*How Well Do We Know Pareto Optimality?*" The Journal of Econ. Education; Vol.22; No:2

Mattelart, A. (2012): Bilgi Toplumunun Tarihi İletişim, İstanbul.

McCombie, J.S.L – Thirwall, A.P. (2004): Essays on Balance of Payments Constraint Growth Routledge.

Meek, R. (1973): Studies in the Labor Theory of Value: From Smith to Ricardo. Lawrance & Wishart, London, 2. Ed.

Meier, G. (Ed) (1976): The Leading Issues in Economic Development Oxford University Press, New York, 3 rd Ed.

Milanovic, B. (2002): "*The Two Faces Of Globalization: Against Globalization As We Know*" http://129.3.20.41/eps/dev/papers/0303/0303007.pdf 2008-11-11

– (2006): "*Global Income Inequality: What It Is and Why It Matters?*" www.un.org/esa/desa/papers/2006/wp26_2006.pdf; DESA Working Paper, No. 26, 2008-05-31

Mohun, S. (1994): "A Re(in)statement Of The Labor Theory Of Value" *Cambridge Journal Of Economics*; 18: 391-412

Mountford, A. (1999): "Trade Dynamics and Endogenous Growth" *Economica*; Vol.66; No.262

Mundell, R.A. (1957): "Transport Costs in Intertational Trade Theory" *The Canadian Journal of Economics and Political Science*; Vol.23; No.3

Nelson, J. (2012): "*Poisoning the Well, or How Economic Theory Damages Moral Imagination*" GDAE Working Paper, No. 12-07

Nelson, R.R. – Phelps, E.S. (1966): "*Investment in Humans, Technological Diffusion, and Economic Growth*", AER, Vol.56; No.1/2; 69-75

Nelson, R.R. – Pack, H. (1999): "*The Asian Miracle and Modern Growth Theory*" The Economic Journal; Vol.109; No.457.

North, C.D. (2002): Kurumlar, Kurumsal Değişim & Ekonomik Performans Sabancı Üniversitesi Yayınları, İstanbul.

Oxley, L. ve diğerleri (2008): "*The Knowledge Economy/Society: The latest example of "Measurement without Theory"?*" Working Paper No. 03-2008; Dept. of Ec., College of Business and Economics Uni. Of Canterbury, New Zealand

Özilgen, M. (2011): Endüstrileşme Sürecinde Bilgi Birikiminin Öyküsü Arkadaş Yayınevi, Ankara.

Özveren, E. (Der.) (2007): Kurumsal İktisat İmge Kitabevi, Ankara.

Parente, S.L. – Prescott, E.C. (1994): "*Barriers to Technology Adoption and Development*"; The Journal of Pol.Ec.; Vol.102; No.2

Pareto, V. (1897): "*The new theories of Economics*" The Journal of Pol. Ec.; Vol.5; No.4; s. 485-502

Patnaik, P. (2013): "*Growth versus Redistribution*" www.networkideas.org/news/aug2013/Growth_Redistribution.pdf, 2013-09-03

– (2004): The New Imperializm IDEAS Int. Conference on The Economics of New Imperialism, 22-24 Jan. 2004

Penrose, E. (1995): The Theory of the Growth of the Firm Oxford University Press.

Phelps, E.S. (Ed.) (1990): Seven Schools Of Macroeconomic Thought. Clarendon Press, Oxford.

Piketty, T. (2014): Kapital Çev. H. Koçak Türkiye Iş Bankası, Kültür Yayınları, Istanbul.

Pissarides, C. (2000): "*Labor Markets & Ec. Growth in MENA Regions*" http://publi.cerdi.org/ed/2005/2005.35.pdf; 2008-09-13

Posner, M.V. (1961): "*International Trade and Technical Change*" Oxford Economic Papers, Vol.13; No.3.

Pons-Vignon, N. (ed.) (2013): Bir Alternatif Var Efil Yayınevi, Ankara.

Prebisch, R. (1959): "*Commercial Policy in the Underdeveloped Countries*" AER, Vol. 49; No.2; May; 251-273

Pyo, H. K. (2001): *"Economic Growth in Korea (1911-1999): A Long -Term Trend and Perspective". Seoul Journal of Economics;* Vol. 14; No. 1

Rawls, J. (2005): A Theory of Justice Harvard University Press.

Rebello, S. (1998): *"The Role of Knowledge & Capital in Ec. Growth"* worldbank. org/research/growth/pdffiles/rebslide.pdf

Ricardo, D. (1990): On The Principles Of Political Economy And Taxation. Cambridge University Press.

Robinson, J. (1962): Economic Philosophy. Penguin Books.

- (1966): The Accumulation Of Capital. Macmillan, London.

- (1972): *"The Second Crisis of Economic Theory"* AER, Vol.62; No.1/2; 1-10

Rockett, K. (2012): *"Perspectives on the Knowledge-based Socity"* http://dx.doi.org/10.5018/economics-ejournal.ja 2012-35

Rodrik, D. (2001): The Global Governance of Trade UNDP, www.undp.org

- (2002): *"After Neoliberalism, What?"* http://ksghome.harvard.edu/~drodrik/after%20neoliberalism.pdf 2008-11-11

- (2005): *"Growth Strategies"* in P. Aghion and S. Durlauf, eds., *Handbook of Economic Growth,* vol. 1A, North-Holland.

- (2009): Tek Ekonomi- Çok Reçete Küreselleşme, Kurumlar ve Ekonomik Büyüme Eflatun Yayınevi, Ankara, Çev. N. Domaniç.

Romer, D. (1996): Advanced Macroeconomics The McGraw-Hill Co., New York.

Romer, P.M. (1986): *"Increasing Returns and Long-run Growth"* Journal of Political Economy, Oct.

- (1990): *"Endogenous Technological Change"* Jour. of Political Economy; Vol.98; October.

- (1993): *"Economic Growth"* in D.R. Henderson (Ed.) *The Fortune Encyclopedia of Economics* Time-Warner Books, N.Y.

- (1994): *"Beyond Classical And Keynesian Macroeconomic Policy". Policy Options*; July-August.

Saint-Paul, G. (2008): *Welfare Effects of Intellectual Property in a North-South Model of Endogenous Growth with Comparative Advantage* www.economics-ejournal.org/economics/journalarticles/2008-5

Salt, J. (1997): *"International Movements of the Highly Skilled"* Occasional Papers No:3, OECD

Samuelson, P.A. (1967): Economics. McGraw-Hill Book Co., New York.

Saygılı, Ş. - Cihan, C. - Yavan, Z.A. (2006): Eğitim ve Sürdürülebilir Büyüme TÜSİAD Büyüme Stratejileri Dizisi, No:7 TÜSİAD-T/2006-06-420, İstanbul.

Scherer, F.M. (1972): Industrial Pricing: Theory and Evidence Rand McNally & Co., Chicago.

Schultz, T.W. (1980): Investing in People. University of California Press

Schumpeter, J.A. (1954): History of Economic Analysis. Oxford Uni. Press, New York.

– (1959): The Theory Of Economic Development Harvard Uni. Press, Cambridge.

– (1970): Capitalism, Socialism And Democracy. Unwin University Books, London.

Semrad, A. 201 "*Modern Secondary Education & Economic Performance*". Uni. Of Munich, Dept. of Ec. Discussion Paper: 2014-47

Sezgin, F. (2007): İslamda Bilim ve Teknik, Cilt:1-5 TÜBA Yayınları No: 14, Ankara.

Shah, A. (2008): "*Poverty Facts and Stats*" www.globalissues.org/article/26/poverty- facts-and-stats; 2008-10-02.

Sherman, H.J. (1991): The Business Cycle Princeton Uni. Press, New Jersey.

Silverberg, G. - Soete, L. (Eds.) (1994): The Economics Of Growth And Technical Change. Edward Elgar Publ.

Simsek, N. (2008): Türkiye'nin Endstri-içi Dış Ticaretinin Analizi Doktora ezi; Beta, İstanbul

Skousen, M. (2003): Modern Iktisadın İnşası (The Making of Modern Economics) Liberte Yayınları, Ankara.

Smith, A. (1976): An Inquiry Into The Nature And Causes Of The Wealth Of Nations, Vol. 1 & 2 Oxford Uni. Press

– (1985): Ulusların Zenginliği Alan Yayıncılık, Istanbul.

Solo, R.A. (1966): "*The Capacity to Assimilate an Advanced Technology*" AER, No:2 (1966): (in Jones, G. 1971)

Solow, R.M. (1956): "*A Contribution to the Theory of Ec. Growth*" The Quarterly Journal of Ec., Vol.70; No.1; 65-94

– (1957): "*Technical Change & the Aggregate Prod. Function*" Review of Ec & Statistics; Vol.39; No.3; 312-320

- (1962): "*Technical Progress, Capital Formation and Economic Growth*", The AER,Vol.52; No.2; 76-86
- (1987): Nobel Prize Lecture, Dec. 8, 1987, Stockholm.
- (1988): Growth Theory: an exposition. Oxford Uni. Press, New York.
- (1994): "*Perspectives on Growth Theory*" The Journal of Ec. Perspectives; Vol.8; No.1

Sörensen, P.B. – Witta-Jacobsen, H.J. (2010): Introducing Advanced Macroeconomics The McGrow-Hill Co.

Soyak, A. (2007): "*Küreselleşme, Teknoloji Politikası, Türkiye*" http://mimoza.marmara.edu.tr/~asoyak/teknoloji-politikasi-sinaimulkiyet(alkan).pdf; 2007-12-15

Sowell, T. (1974): Classical Economics Reconsidered. Princeton Uni. Press, Princeton, New Jersey.

Spolaore, E. – Wacziarg, R. (2013): "*Long-term Barriers to Economic Development*" NEBR Working Paper No: 19361

Stanfield, J.R. (1979): Economic Thought and Social Change. Southern Illinois Uni. Press.

Stiglitz, E. J. (2002): Küreselleşme: Büyük Hayal Kırıklığı Plan B, İstanbul.

Talalay, M. – Farrands, C. – Tooze, R. (Eds.) (1997): Technology, Culture and Competitiveness. Routledge, London.

Tarascio, V.J. (1971): "*Keynes on the Sources of Economic Growth*" The Journal of Economic History, Vol.31; No.2

Taymaz, E. – Suiçmez, H. (2005): "*Türkiye'de Verimlilik, Büyüme ve Kriz*" TEK Tartışma Metni, 2005/4; www.tek.org.tr/dosyalar/TAYMAZ-SUICMEZ.pdf; 2007-09-12

Temple, J. (1999): "*The New Growth Evidence*" Journal of Economic Literature; Vol.37; No.1.

Tezel, Y.S. (2003): Iktisadi Büyüme Imaj Kitabevi, Ankara.

Tobin, J. (1996): Full Employment and Growth Edward Elgar, U.K.

Todaro, M.P. – Smith, S.C. (2003): Economic Development Addison-Wesley Series.

Taban, S. (2010): İçsel Büyüme Modelleri ve Türkiye Ekin, Bursa.

Toffler, A. (1992): Yeni Güçler - Yeni Şoklar (Powershift) Altın Kitaplar, İstanbul

Türkcan, E. (1981): Teknolojinin Ekonomi Politiği AİTİA Yayın No: 151, Ankara.

Türkmen, F. (2002): Eğitimin Ekonomik ve Sosyal Faydaları ve Türkiye'de Eğitim-Ekonomik Büyüme İlişkisinin Araştırılması. DPT, Uzmanlık tezi, Yayın No:2655.

Ulutan, B. (1978): İktisadi Doktrinler Tarihi. Ötüken Neşriyat, İstanbul.

Ünsal, E.M. (2007): İktisadi Büyüme İmaj Yayıncılık, Ankara.

Vásques, I. (Ed.) (2003): Kapitalizm ve Küresel Refah Liberte Yayınları, Ankara.

Vernon, R. (1966): "International Investment and International Trade in the Product Cycle"; The Quarterly Journal of Economics; Vol.80; No.2, 190-207.

Yang, Xiaokai (2003): Economic Development & the Division of Labour Blackwell Publ.

Wacziarg, R. (2002): "Review of Easterly's The Elusive Quest for Growth." Journal of Ec. Literature, Vo. XL, Sept.2002

Weintraub, S. (Ed.) (1977): Modern Economic Thought. University Of Pennsylvania Press.

Whitaker, J.K. (2003): "Henry George and Classical Growth Theory" American Journal of Economics & Sociology, Vol:60, Issue:1, Oct.-2003

Yeldan, E. (2010): İktisadi Büyüme ve Bölüşüm Teorileri Efil Yayınevi, Ankara.

Yeldan, E. – et. al. (2012): "Orta Gelir Tuzağından Çıkış: Hangi Türkiye" TÜRKONFED Raporu, Cilt-1, Sektörel Analiz.

– (2013): "Orta Gelir Tuzağından Çıkış: Hangi Türkiye" TÜRKONFED Raporu, Cilt-2, Çözüm önerileri.

Yıldırım, N. (1973): Neoklasik İktisadın Teknolojik Gelişme Yaklaşımı, A.Ü., SBF Yayını No:367

Yıldırım, K. – Karaman, D. (2003): Makroekonomi Eğitim. Sağlık ve Bilimsel Araştırma Çalışmaları Vakfı, Yayın no: 145

Yücel, İ.H. (1997): "Bilim-teknoloji Politikaları ve 21. Yüzyıl'ın Toplumu" http://ekutup.dpt.gov.tr/bilim/yucelih/biltek.pdf 2008-05-09

Yıldırım, K. – Karaman, D. (2003): Makroekonomi Eğitim, Sağlık ve Bilimsel Araştırma Çalışmaları Vakfı, Yayın No. 145; 3. Basım.

Young, A. (1998): "Growth Without Scale Effects" The Jour.of Political Economy, Vol.106; No.1

Wade, R.H. (2014): "The Piketty phenomenon & the future of inequality" Real-world Economics Review, Issue:69 www.paecon.net/PAEReview/issue69/whole69.pdf 2014-10-08

Weber, M. (2008): Protestan Ahlâkı ve Kapitalizmin Ruhu Ayraç Kitabevi, Ankara, 5. Baskı

Weeks, J. (2008): *"A Critique of Neoclassical Macroeconomics"* 1989 Revised as False Paradigm, part 1 & part 2 http://jweeks.org/Economic_Theory.html, 2010-07-10

– (2014): Economics of the 1 % Anthem Press, London.

Weintraub, S. (Ed.) (1977): Modern Economic Thought. University Of Pennsylvania Press.

Zambelli, S. (2010): Computable, Constructive and Behavioural Economic Dynamics, Routledge.

Other Sources:

<u>BSB</u> (2005): *"On Economic and Social Life in Turkey in Early 2005"* www.bagimsizsosyalbilimciler.org/; 2005-07-17 Bağımsız Sosyal Bilimciler.

<u>DIE-TÜIK</u> (2004): Türkiye İstatistik Yıllığı www.die.gov.tr/yillik/yillik_2004.pdf; 2005-07-16

<u>DPT</u>

– (2001): *"Ortaöğretim: Genel eğitim, meslek eğitimi, teknik eğitim"*: Özel İhtisas Komisyonu Raporu.

– (2005): 8. Beş Yıllık Kalkınma Planı 2001-2005 http://ekutup.dpt.gov.tr/program/2005.pdf: 2005-07-23

– (2005): Türkiye Ekonomisinde Sermaye Birikimi, Verimlilik ve Büyüme: 1972-2003. Ankara, Yayın No. 2686

<u>EU</u> (2005): Science and Technology in Europe. Eurostat.

<u>ILO</u> (2005): www.ilo.org/public/english/employment/strat/kilm/trends.htm#-figure%201a; 2005-07-07

<u>IMF</u> (2002): Balance of Payments Statistics Yearbook New York.

<u>MPM</u> (2004): Verimlilik Raporu-3, Ankara.

<u>OECD</u> (1993): Intra-Firm Trade Trade Policy Issues-1, Paris.

– (2000): Investing in Education

– (2004)-a: Main Science and Technology Indicators Paris.

– (2004)-b: Ecnomic Outlook. No. 76, Aralık.

- (2005): Employment Outlook www.oecd.org/els/employmentoutlook
- (2014): OECD, Labor market outcomes www.oecd.org/els/emp/oecdlabourmarketoutcomes-unemployment.htm, 2014-07-22

<u>PRB</u> (2004): World Population Data Sheet www.prb.org/pdf04/04WorldDataSheet_Eng.pdf Population Reference Bureau.

<u>TEK</u> (2003): Büyüme Stratejileri, Türkiye Ekonomi Kurumu Ankara.

<u>TÜSIAD</u> (2004): Türkiye Ekonomisi, Aralık-2004 Yayın No. TÜSİAD-T/2004-12-384

<u>UNCTAD</u> (2014): Trade and Development Report UNCTAD/TDR/2014.

<u>WORLD BANK</u> (2000): The Quality of Growth Oxford University Press.

- (2002): Human Development Report http://www.worldbank.org/
- (2003): Global Economic Prospects http://www.worldbank.org/
- (2004): World Development Indicators http://www.worldbank.org/
- (2008): Using Knowledge to Improve Development Effectiveness http://web.worldbank.org/WBSITE/EXTERNAL/EXTOED/EXTECOSECWOR/0,,-contentMDK:21855174~menuPK:5249468~pagePK:64829573~piPK:64829550~theSitePK:5249459,00.html; 2008-09-19

www.ingramcontent.com/pod-product-compliance
Ingram Content Group UK Ltd.
Pitfield, Milton Keynes, MK11 3LW, UK
UKHW021829210426
5322IPUK00004B/104